Drug Trafficking and Protected Areas in Mexico

José Luis Carpio-Domínguez •
Jesús Ignacio Castro-Salazar •
César Hernández-Mier

Drug Trafficking and Protected Areas in Mexico

A Green Criminological Perspective on Organised Crime, Environment, and Public Security

Springer

José Luis Carpio-Domínguez
Universidad Autónoma de San Luis Potosí
San Luis Potosí, Mexico

César Hernández-Mier
Universidad Autónoma de San Luis Potosí
San Luis Potosí, Mexico

Jesús Ignacio Castro-Salazar
TecNM/Instituto Tecnológico Superior
de Abasolo
Guanajuato, Mexico

ISBN 978-3-032-18217-3 ISBN 978-3-032-18218-0 (eBook)
https://doi.org/10.1007/978-3-032-18218-0

This Springer imprint is published by the registered company Springer Nature Switzerland AG
The registered company address is: Gewerbestrasse 11, 6330 Cham, Switzerland

Foreword

There is an inextricable link between plant-based narcotics and the environment and thus conservation. Yet despite the likely impacts of the illegal drug economy on nature, few studies have documented the consequences of growing crops for narcotic drugs on the environment. Those studies that have been done have tended to focus on the impacts of chemical eradication programmes, pollution generated by drug manufacturing, or—to some degree—deforestation. The book *Drug trafficking and Protected Areas in México: A Green Criminological Perspective on Organised Crime, Environment, and Public Security* provides an altogether different approach to this important topic.

Providing a robust history, geographical overview, and legislative analysis of natural protected areas in Mexico, this book starts from a green criminological foundation by placing the environment at the centre of this conflict. This contextualises the necessary background on the governance and enforcement of laws in these conservation spaces as well as weaving in understanding of the human relationship to these areas. The authors then go on to detail Mexico's long and complex drug trafficking history—from prohibition and Chinese control of the market to the rise of Mexico's pax narcotica and its own "war on drugs". All the while, the impact of drug trafficking on the environment in Mexico is highlighted as is how this has affected various groups of people.

The final part of the book richly details the current situation. The locations of organised crime groups and natural protected areas are mapped, showing the extensive convergence of the two throughout all of Mexico. The strength of the book is that not only does it provide such granular technical detail, but this is complimented with the powerful lived experiences of the people confronted with these crimes. A key takeaway is that drug trafficking across Mexico is variable; each region requires tailored responses to protect the environment and people from the violence.

This valuable original contribution to both green criminology and drug scholarship sets an excellent example of interdisciplinarity and mixed methods research design. History, geography, law, political science, natural sciences, and critical and green criminology combine to document the threats to people and planet posed by powerful criminal groups that recruit and exploit marginalised people and the environment. The book is a standard setter for multidisciplinary projects that generate impactful data to inform policy and better protect the environment.

United Nations Office on Drugs and Crime, United Nations
Vienna, Austria

Tanya Wyatt

Acknowledgements

The authors would like to thank the inspectors and directors (women and men) of protected areas in Mexico (CONANP) for all their support and accessibility during the interviews. They would also like to thank the staff at Springer, in particular Anna Goodlett for the help and support.

Competing Interests The authors have no competing interests to declare that are relevant to the content of this manuscript.

Contents

Contents

About the Authors

José Luis Carpio-Domínguez, PhD He is a Graduate in Criminology from the Northwestern University of Chihuahua, Mexico. He holds a Master in Criminology and Forensic Sciences from the Universidad Autónoma de Tamaulipas, Mexico. He obtained his PhD in Social Sciences with a focus on Sustainable Development from the Social Research Institute of the Universidad Autónoma de Nuevo León, Mexico. He is a research professor in the Criminology Department of the Law Faculty at Universidad Autónoma de San Luis Potosí. He is a member of the National System of Researchers of Mexico (SECIHTI-SNII-I), a member of the Mexican Society of Criminology, a member of the Thematic Network of Forensic Sciences from SECIHTI (UNAM), and a member of the Green Criminology Specialist Group of the International Union for Conservation of Nature (IUCN). He develops scientific research in green criminology with topics related to different forms of environmental crime.

Jesús Ignacio Castro-Salazar, PhD. He is an Environmental Engineer from the University of Guanajuato, a Master in Environmental Administration from El Colegio de la Frontera Norte, and a Doctor in Social Sciences with a focus on Sustainable Development from the Universidad Autónoma de Nuevo León, Mexico. He has worked as an inspector in the Federal Environmental Protection Agency. He collaborates on socio-environmental and energy projects applied in rural and indigenous communities in Mexico. He is a member of the National System of Researchers of Mexico (SECIHTI-SNII-I) and a member of the Green Criminology Specialist Group of the International Union for Conservation of Nature. He is a Professor-Researcher at the Instituto Tecnológico Nacional de México / Instituto Tecnológico Superior de Abasolo. His main lines of research are environmental public policy and administration; sociology and environmental legislation; green criminology.

César Hernández-Mier MSc. He obtained his Doctor of Medicine from the Autonomous University of Querétaro; he holds a specialisation in Forensic Medicine from the Autonomous University of the State of Mexico, based at the Attorney General's Office of the State of Mexico; he holds a Master's Degree in Forensic Medicine from the University of Valencia in Spain. Currently, he is a Research

Professor in the Criminology Department of the Law Faculty at the Autonomous University of San Luis Potosí, and carries out teaching and research activities in criminology, human identification, and human rights.

Chapter 1
Introduction

The presence of drug traffickers in protected areas in Mexico is a phenomenon at the intersection of public security and environmental conservation that has not yet been recognised as an issue of interest in academia or public policy. Although environmental conservation and public security may seem like socio-environmental, economic, political, and disciplinary phenomena that are apparently unrelated, the presence of criminal drug trafficking groups that threaten public security and environmental conservation always emerges when analysing environmental conservation in Mexico, particularly in protected areas.

Several studies in Latin America have reported on the social and environmental harms caused by organised crime within protected areas (Devine et al., 2021a, b; Felbab-Brown, 2018; Wrathhall, 2020) and have highlighted the importance of promoting effective government action to enforce the law. However, over the years, this phenomenon appears to be spreading geographically to new protected areas as criminal groups grow in dominance and strength.

Drug trafficking, as an entity that undermines the rule of law in Mexico, did not arise spontaneously, but rather developed and adapted in response to socio-historical, political, cultural, and economic phenomena. In addition, Mexico's environmental characteristics, such as the orography of mountain ranges such as the Sierra Madre Occidental and the Sierra Madre Oriental, the Neovolcanic Axis in Central Mexico, the Sierra Madre de Chiapas and the Sierra Madre del Sur, result in highly fertile soil, water availability, and remoteness from urban centres, which were fundamental for the planting, harvesting, and trafficking of drugs such as poppies and marijuana, as well as for the consolidation of criminal groups with expertise in drug production and trafficking.

Protected areas are geographical territories where the aim is to curb the impact of human activities on the environment, and these areas often coincide with strategic territories for drug cultivation, harvesting, and trafficking in Mexico. Protected areas have even been established in territories already occupied for drug production activities, mainly on the continental territory, while several protected areas in marine

J. L. Carpio-Domínguez et al., *Drug Trafficking and Protected Areas in Mexico*,
https://doi.org/10.1007/978-3-032-18218-0_1

territory coincide with the routes used by criminal groups to transport drugs, firearms, and migrants by boat and motorboat.

The presence of criminal drug trafficking groups in protected natural areas is multifactorial; various social, cultural, environmental, political, and economic phenomena have been fundamental to the establishment of these criminal groups in areas dedicated to environmental conservation.

The presence of drug traffickers in protected areas has been ambivalent. They pose a serious threat to the environmental conservation strategies implemented by the State, as well as to the security of rural and indigenous communities and public servants responsible for environmental conservation. Nevertheless, it has been documented that the potential threat of drug traffickers' presence in these areas has deterred poaching, illegal logging, and deforestation (see Wyatt, 2015; Carpio-Domínguez, 2021; 2024).

This book provides an interdisciplinary exploration of the convergence between drug trafficking and environmental conservation within Mexico's protected areas. Bridging environmental studies, green criminology, and political geography, it examines how criminal groups exploit these territories for the production, processing, and transit of narcotics, posing significant challenges to public security, ecological preservation, and social justice.

Through a critical analysis of geographical, social, cultural, and governmental factors, the book addresses the historical context of drug trafficking in Mexico, the impact of policies such as the Mexican War on Drugs, and the emergence of new illicit economic markets in protected areas. By incorporating testimonies from public servants and inspectors, it highlights from a green criminological perspective, the complexities of managing environmental conservation in regions under the influence of organised crime.

The green criminological perspective serves in this book not only as an epistemological framework, but also to thread together the social, economic, and environmental phenomena related to the influence and harms caused by drug trafficking in Mexico's protected areas.

Drug Trafficking and Protected Areas in Mexico is a timely contribution that emphasises the need for integrative approaches to understanding the nexus of public security and environmental conservation. It challenges traditional disciplinary boundaries by foregrounding the intricate links between ecological sustainability, governance, and the socio-economic dynamics of organised crime.

The main limitation of this book is that it only considers federal protected areas and does not consider those under the jurisdiction of each state (state protected areas), which possibly face similar and perhaps more serious problems than those documented in this book, but which remain among the pending research in the study of drug trafficking and Mexico's protected areas.

This book not only seeks to document drug trafficking and the activities of criminal groups in protected areas, but also to visualise the experiences of environmental inspectors and directors that have remained only in the individual and collective memory of these people who dedicate their lives and professions to environmental conservation in Mexico through public service. The authors hope that this book will bring justice by recognising their work, which not only benefits species and habitats within protected areas, but also cares for the future of Mexico and our planet.

We not only protect nature, but we also protect human life.

Environmental inspector from CONANP.

Part I
Protected Areas in Mexico

Overview

This section provides the basis for understanding the critical role of protected areas in Mexico's environmental conservation efforts and their vulnerability to external pressures, including organised crime. This section contextualises the relevance of protected areas in Mexico, giving a brief overview of how the protected areas were decreed and their current geographical distribution and territorial extension.

These chapters analyse the environmental services provided by protected areas and their cultural and economic relationship with indigenous and rural communities. This section also analyses the institutional efforts to guarantee environmental conservation in the country and discusses the relevance of protected areas for social development and the main threats to environmental conservation in these territories. Through an analysis of the environmental legislation and the institutional structure for its legal protection.

The second chapter explores the development of protected areas in Mexico, examining the historical, political, and institutional factors that shaped their establishment and evolution, and highlights how these areas became a cornerstone of the country's conservation strategy. The geographic distribution of protected areas is mapped to contextualise with ecological characteristics but also with historical and social development.

The third chapter analyses the geographical distribution and ecological significance of protected areas, emphasising their role in biodiversity conservation and the socio-economic relationships they sustain with indigenous and rural communities. Highlighting the importance of natural protected areas as "strategic territories" for environmental conservation, but also for illegal activities.

The fourth chapter presents and analyses the types of governance in Mexico's protected areas and explores the governmental apparatus and its limitations to achieve the protected areas' objectives. This allows to understand the limitations and strengths of the State to face the challenges in Mexico's protected areas.

Chapter 2
Mexican History of Protected Areas

Introduction

Around the world—and in Mexico—protected areas are one of the best legal strategies and instruments for addressing the challenge of environmental conservation (Dudley, 2008; Ervin et al., 2010). Protected areas are important for environmental conservation, and host ecosystems, habitats and a great biodiversity of flora and fauna species and genetic resources. Protected areas are also cultural sites where people live and allow society to contemplate and understand the natural world through visits and tourism, as well as providing environmental services (Rivas Toledo, 2006; Eagles, 2012). Despite their importance, protected areas face challenges, both within and beyond their boundaries, with (human) activities that contribute to the loss of ecosystems and biodiversity (International Union for Conservation of Nature [IUCN], 2005; Correa, 2016).

There are different scales and categories of protected areas: international, national, regional, state, and local. In Mexico, there are various categories of protected areas, ranging from those voluntarily designated for conservation to those proposed by state and municipal governments. However, protected areas administered by the federal government are the most recognised within national and international environmental policy. Mexican federal protected areas adopt the international categorisation system and policy proposed by the International Union for Conservation of Nature (IUCN). In Mexico, protected areas are known as natural protected areas, but have the same general objective proposed at the international level, which is to conserve nature, biodiversity, cultural resources, and environmental services.

The first protected area in the country was established in 1917, and by the end of 2024, 232 had been decreed throughout the national territory, increasing not only in number but also in extension and diversity of categories. To improve management and address potential social conflicts within and on the periphery of protected areas,

J. L. Carpio-Domínguez et al., *Drug Trafficking and Protected Areas in Mexico*,
https://doi.org/10.1007/978-3-032-18218-0_2

it is important to understand and generate knowledge about the evolution of protected areas (IUCN, 2005; Brenner, 2010). This chapter describes the historical evolution in the number, categories, and territorial extension of protected areas in Mexico from the first decree until 2024.

About Protected Areas, from International to National

Internationally protected areas are considered strategies for conserving biodiversity and ecosystems, preventing species extinction, and protecting the environmental resources and services that these areas provide to society (Murillo Flores & Orozco Alvarado, 2006; Bezaury-Creel & Gutiérrez Carbonell, 2009). One of the most widely recognised official definitions of protected areas is that presented by the International Union for Conservation of Nature (IUCN), which conceptualises them as defined, recognised, dedicated, and managed geographical areas, through legal or other effective means, to achieve the long-term conservation of nature, including ecosystem services and associated cultural values (Dudley, 2008; The World Bank, 2010, p. 1).

Protected areas are considered key components of national and international conservation strategies used to protect and maintain biomes, ecosystems, and wildlife species. Protected areas are seen as a hope for preventing the extinction of species, especially those considered threatened or endemic,[1] as they provide spaces to prevent the loss of the world's natural capacity to sustain different forms of life, including humans. Protected areas also act as carbon sinks and buffer zones against climate change (Secretariat of the Convention on Biological Diversity, 2008; Dudley, 2008; The World Bank, 2010; Costa Pellizzaro et al., 2015).

Protected areas are not only ecosystems, habitats, and wildlife species, they are also cultural places where people live and where the natural world can be contemplated and understood through activities such as tourism, scientific research, and environmental education, providing social, economic, and cultural benefits (Rivas Toledo, 2006; The World Bank, 2010; Eagles et al., 2012). Covering large areas of territories, some protected areas are home to indigenous peoples and rural, Afro-descendant or mestizo communities, and influence towns and cities located on the peripheries of these areas (Rivas Toledo, 2006; Costa Pellizzaro et al., 2015).

The natural resources protected within these areas are often essential to ensuring healthy communities and are the economic engine for many localities, as millions of people depend on protected areas for their subsistence. Some benefit directly through the consumption of food produced or obtained within or near protected areas, while others benefit indirectly through employment and economic income

[1] National endemic species are unique to a specific geographical region, their geographical distribution is limited, and their population size is usually small, which increases their vulnerability to extinction (Kraus et al., 2023; Moya et al., 2025).

(Secretariat of the Convention on Biological Diversity, 2008; The World Bank, 2010; Food and Agriculture Organization of the United Nations [FAO], 2014).

In recent years, protected areas have become part of the Sustainable Development Goals (SDG's) worldwide. For example, they are included in SDG's 3, 6, 11, 13, 14, and 15 for their contribution to biodiversity conservation, cultural resources, and contributing to sustainable development, particularly at the local level (United Nations, 2024). In the management of protected areas, it is essential to understand not only the ecosystemic dimension but also their social and cultural dimensions and the interests and needs of the population that depends directly or indirectly on these resources (Gravez et al., 2011).

The IUCN has set multiple objectives for protected areas, including: (a) conserving the composition, structure, function, and evolutionary potential of biodiversity; (b) contributing to regional conservation strategies; (c) maintaining landscape or habitat diversity; (d) be of sufficient size to ensure the integrity and long-term maintenance of specific conservation objectives or be capable of being enlarged to achieve this purpose; (e) maintain their values in perpetuity; (f) operate under a management plan and a monitoring and evaluation programme that supports adaptive management; and g) have a clear and equitable governance system (Dudley, 2008, p.6).

The world's first protected area was established in 1872, Yellowstone National Park in the United States of America (Costa Pellizzaro et al., 2015). Until 2024, there were approximately 261,766 officially recognised protected areas worldwide, covering more than 15.0% of the planet's surface area and 7.4% of the world's oceans (IUCN, 2025). However, until the mid-twentieth century, there was no international standardisation of the criteria for defining the administrative nomenclature and objectives of these areas, and therefore each country adopted its own terminology, leading to conflicts when these areas were established and analysed at the global level (Costa Pellizzaro et al., 2015).

Over the years, based on a global strategy and with the aim of reducing international conflicts associated with the heterogeneity of concepts, policies, and strategies for the establishment and management of protected areas, in 1948, at the International Conference organised by the United Nations Educational, Scientific and Cultural Organisation (UNESCO) held in Fontainebleau, France, the International Union for the Protection of Nature (IUNP) was created, which in 1955 changed its name to the International Union for Conservation of Nature (IUCN), and included the establishment of internationally recognised guidelines for the implementation and administration of protected areas (Behrman, 1974:21; Costa Pellizzaro et al., 2015).

At the 1992 World Congress of National Parks and Protected Areas, with the aim of standardising the management of protected areas internationally, the IUCN established an international system of management categories based on aspects such as the level of intervention and the naturalness of the environments.

This system was subsequently revised and updated in 2008. The categories are: Category Ia: Strict Nature Reserve, Category Ib: Wilderness Area (Ib), Category II: National Park, Category III: Natural Monument, Category IV: Species/Habitat

Management Area, Category V: Protected Landscape or Marine Area, and Category VI: Protected Area with Sustainable Use of Natural Resources (Bishop et al., 2004:64; Naughton-Treves et al., 2005; Dudley, 2008; United Nations Environmental Programme-World Conservation Monitoring Centre [UNEP-WCMC], 2018).

The system developed by the IUCN is not mandatory, and each country can choose whether to incorporate the guidelines into its legislation governing protected areas, which makes it difficult to achieve the objectives given the wide variety of positions adopted by many countries (Costa Pellizzaro et al., 2015). Mexico is one of the countries that uses the IUCN categories, and having a significant influence (Bishop et al., 2004:64).

In Mexico, there are protected areas, which are one of the Mexican government's responses to the conservation and protection of species and ecosystems.(Carpio-Domínguez, 2021). Mexico is recognised worldwide as one of the most megadiverse countries due to the biological diversity that exists within its national territory. Mexico stands out internationally as a nation with a large number and diversity of animals and plants, accounting for almost 70% of the world's species diversity, and for its high rate of endemism,[2] especially in the southern and south-eastern regions of the country and in northern states such as Baja California and Baja California Sur (Comisión Nacional de Áreas Naturales Protegidas [CONANP], 2012; Sarukhán et al., 2009).

The number of species present within the national territory has positioned it globally in 11th place for its diversity of birds, fifth place for vascular plants, fifth place for amphibians, third place for mammals, and second place for reptiles (Comisión Nacional para el Conocimiento y Uso de la Biodiversidad [CONABIO], 2023a). Of the total number of vertebrates inhabiting Mexico, 20.0% are endemic (found only within Mexican territory), of which 46.0% of reptiles are endemic, 30.0% of mammals, 48.0% of amphibians, and 11.0% of birds; and it is estimated that between 50.0% and 60.0% of the vascular flora known in the country are endemic (Sarukhán et al., 2009).

In Mexico, protected areas are officially recognised as land or water areas within the national territory where the original environments have not been significantly altered by human activity or that require preservation and restoration and are subject to conservation regimes, over which the nation exercises its sovereignty and jurisdiction (Article 3, Section II, of the Ley General del Equilibrio Ecológico y Protección al Ambiente [LGEEPA], 2024).

The decree and administration of protected areas in the country have among their objectives: (a) to preserve natural environments representative of different biogeographical and ecological regions and the most fragile ecosystems in order to ensure the balance and continuity of evolutionary and ecological processes; (b) to safeguard the genetic diversity of wild species on which evolutionary continuity depends; (c) ensuring the preservation and sustainable use of ecosystems, their elements, and their functions; (d) generating, rescuing, and disseminating traditional or

[2] Endemism is a concept used in biology to identify native taxa or biological groups with a restricted distribution area, i.e. the distribution of a taxon is limited to a small geographical area and, therefore, it can only be found naturally in that location (Noguera-Urbano, 2017).

new knowledge, practices, and technologies; and (e) protecting the natural environments of areas, monuments, and archaeological, historical, and artistic remains, among others (Article 45 of the LGEEPA, 2024).

There are various categories of protected areas in the country, depending on the authority that decrees and administers them: by the federal government, by state governments, by municipal governments, or those voluntarily designated for conservation by governments or communities and communal lands, including private territories.

Protected areas designated and administered by the federal government are the most politically and environmentally significant at the national and global levels because protected areas are internationally recognised. The categories of federal protected areas coincide with the categories recognised by the IUCN, with the following existing in the country: Flora and Fauna Protection Areas (APFF its Spanish abbreviation), Natural Resource Protection Areas (APRN), Natural Monuments (MN), National Parks (PN), Biosphere Reserves (RB), and Sanctuaries (SANT) (Íñiguez Dávalos et al., 2014; CONABIO, 2023b; Article 46 of the LGEEPA, 2024).

The Comisión Nacional de Áreas Naturales Protegidas[3] (CONANP) (National Commission for Protected Natural Areas) (2024) broadly defines the characteristics and importance of each of these categories of areas as follows:

(a) Flora and Fauna Protection Areas (APFF): These are areas that contain habitats on whose balance and preservation the existence, transformation, and development of wild flora and fauna species depend.
(b) Natural Resources Protection Areas (APRN): These are areas designated for the preservation and protection of soil, hydrographic basins, water and, in general, natural resources located in forest terrains. These areas include reserves, forest areas, protection zones for rivers, lakes, lagoons, springs and bodies of water considered national waters, particularly those that supply water to the population.
(c) National Parks (NP): Areas that are significant for their landscape aesthetics, scientific, educational and recreational value, historical value, flora and fauna, suitability for tourism development, or other reasons of general interest.
(d) Natural Monuments (MN): These are areas containing natural elements, consisting of natural sites or objects which, due to their unique character, aesthetic interest, historical or scientific value, are incorporated into a regime of absolute protection. These areas do not have the diversity of ecosystems or the surface area necessary to be included in other categories.
(e) Biosphere Reserves (RB): Areas containing ecosystems that have not been significantly altered by humans, or that are to be preserved and restored, and which are home to species that are representative of the country, endemic, threatened, or endangered.

[3] The National Commission for Protected Natural Areas (CONANP) is a Mexican federal government agency responsible for the administration and preservation of protected natural areas within the country.

(f) Sanctuaries (SANT): Areas characterised by considerable richness in flora or fauna, or by the presence of species, subspecies, or habitats with restricted distribution. These areas include canyons, meadows, relict areas, caves, caverns, cenotes, coves, or other topographical or geographical features that need to be preserved or protected.

Each of these categories of protected areas has its own characteristics. The restrictions within these protected areas are related to their creation and administration and are consistent with their biological, physical, and socio-economic elements, as well as their category and management plans. The purpose of these restrictions is to fulfil the objectives for which they were created, issues that will be addressed in the following chapters. These issues will be developed in the next two chapters.

Protected areas are even part of national environmental policy, as these areas are recognised in the plans and programmes of various government administrations, such as the National Development Plan (PND), a document that sets out the Mexican government's objectives and strategies for the country's development. Although the most recent National Development Plan 2019–2024 did not include protected areas among its objectives, strategies, or actions, or within the content of the document (Gobierno de México, 2019), its predecessor, the 2013–2018 National Development Plan, mentioned the importance of ensuring that natural resources continue to provide environmental services to the population, including the 12% of the national territory designated as protected areas (Gobierno de México, 2013).

The National Development Plan 2019–2024 gives rise to sectoral programmes, such as the Sectoral Programme for the Environment and Natural Resources (PROMARNAT) 2020–2024 (Secretaría de Medio Ambiente y Recursos Naturales [SEMARNAT], 2020), which establishes as a priority objective the promotion of the conservation, protection, restoration, and sustainable use of ecosystems and their biodiversity, setting out the following specific actions: (1) consolidating and promoting protected areas, (2) promoting the effective management of protected areas through the participation of vulnerable groups, such as women, young people, and indigenous and Afro-Mexican peoples and communities.

The Evolution of Protected Areas in Mexico

The decrees and administration of protected areas in Mexico have followed international trends, initially influenced by the American conservationist model, given that the first protected areas were decreed under the categories of national park management (Martínez et al., 2016). The first protected area in Mexico was established on 12 November 1917. It was a terrestrial national park located in Mexico City, within the Central and Neovolcanic Region, and was named "Desierto de los Leones" (Lions' Desert). It was not until 1935 that a new protected area was decreed, almost 20 years later.

Since the first protected area in 1917 until the end of 2024, many protected areas under federal jurisdiction with different categories or protection statuses have been decreed in the country. Mexico went from having one protected area in 1917 to 232 protected areas in 2024 (Fig. 2.1). During the period from the first protected area in 1917 to the end of 2024, in 51.9% (n = 56) of the years no protected areas were designated, while in 48.1% (n = 52) of the years at least one protected area was designated. However, between 1917 and the last year in which a protected area was designated in the country (2024), an average of two protected areas per year was designated. The years 1986 (n = 18), 2023 (n = 19), and 2024 (n = 27) were the years when the highest number of protected areas were decreed in Mexico (Fig. 2.1).

When considering 20-year periods in which protected areas were decreed, in the first 20 years since the first protected area in 1917 and at the end of 1936, a total of 13 protected areas were decreed. In the following 20 years, at the end of 1956, the cumulative number of protected areas increased by 284.6% (n = 50). Subsequently, only five new protected areas were added until 1976, bringing the total to 55. By the end of 1996, protected areas had increased by 134.5%, reaching a total of 129, then in 2016, protected areas increased by 41.1% compared to the previous period, to 182, and by the end of 2024, protected areas increased by 27.5%, to 232 (Table 2.1).

An analysis conducted between 1985 and 1997 published by the IUCN and the WCMC of the United Nations Environment Programme (UNEP) revealed that the largest percentage increase in protected areas occurred in Mesoamerica (comprising Mexico, Central America, and the Caribbean) and Western/Mediterranean Europe (Western Europe and the European side of the Mediterranean), 10.4% and 10.3%, respectively (Naughton-Treves et al., 2005).

Another study covering the period from 1980 to 2005 found that the area of land under legal protection increased exponentially, with more than 100,000 protected areas covering 17.1 million square kilometres (km^2) (11.5% of the planet's land surface) being protected (Naughton-Treves et al., 2005). Meanwhile, in 2010, it was

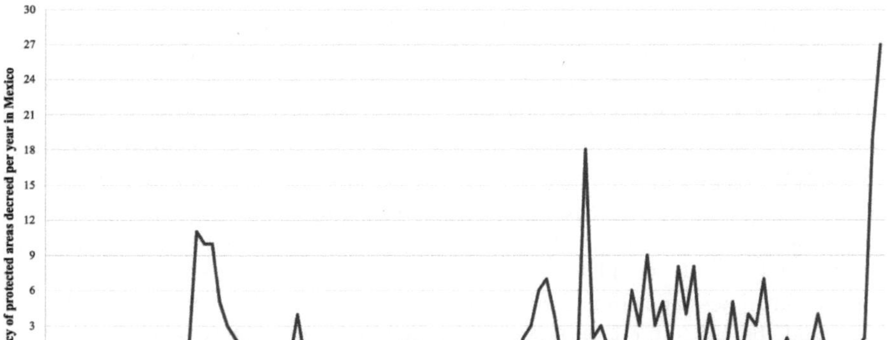

Fig. 2.1 Frequency of protected areas decreed per year in Mexico, 1917–2024. **Source:** the authors with data from Comisión Nacional de Áreas Naturales Protegidas (2025)

Table 2.1 Evolution of protected areas in Mexico by 20-year periods, 1917–2024

Period	Protected areas	% increase
1917–1936	13	First 20 years
1937–1956	50	284.6
1957–1976	55	10.0
1977–1996	129	134.5
1997–2016	182	41.1
2017–2024	232	27.5

Source: the authors with data from Comisión Nacional de Áreas Naturales Protegidas (2025)

vvestimated that protected areas covered around 13.9% of the planet's land surface, not including unofficial protected areas (such as community reserves, sacred natural sites, and indigenous peoples' lands) (The World Bank, 2010).

The first substantial increase in protected area decrees in Mexico in 1986 was driven by an environmental policy within the country during those years. In 1984, for the first time, the federal government presented a National Ecology Programme (1984–1988), which explicitly introduced criteria for the control and prevention of environmental deterioration and the sustainable use of natural resources with a long-term vision (Secretaría de Desarrollo Urbano y Ecología [SEDUE], 1984).

The National Ecology Programme (1984–1988) recognised that natural parks and reserves had not formed a true system of protected areas, did not have defined objectives, and did not adequately represent the country's ecological regions, as important ecosystems were not protected. Chapter 2 of the programme, in its objective 2, was the first official document in the country to set out two specific objectives: (1) to integrate a system of protected areas that would enable their operation and control and (2) to increase the area of protected natural areas by diversifying the representativeness of the country's ecological regions. These official declarations led to an increase in the number of protected area decrees within the country 2 years after the programme's publication. Prior to the programme's entry into effect, there were 77 protected areas until 1985, and by 1986 there were already 95.

The other 2 years with the highest number of protected area decrees in Mexico, 2023 and 2024, were driven by national and international commitments. First, the 2020–2024 Sectoral Programme for the Environment and Natural Resources (SEMARNAT, 2020), published in June 2020, states that in order to achieve one of the country's priority objectives related to promoting the conservation, protection, restoration, and sustainable use of ecosystems and biodiversity, specific actions must be taken to consolidate and promote protected areas, giving priority to the representativeness and connectivity of ecosystems, the conservation of priority species, and the biocultural heritage of the human communities that inhabit them (SEMARNAT, 2020).

Second, the increases in protected areas in 2023 and 2024 were related to the country's commitments made at the Conference of the Parties to the Convention on Biological Diversity (COP 15) in Montreal, Canada, in 2022. At COP 15, 23 goals were set for the leaders and official representatives of the 196 member countries, focused on establishing a plan to stop biodiversity loss by 2030 and restore it by 2050.

Goal 3 stipulates "Ensure and enable the effective conservation and management of at least 30% of terrestrial, inland water, coastal and marine areas, especially areas of particular importance for biodiversity and ecosystem functions and services, by 2030, through ecologically representative protected area systems..." (United Nations Environment Programme [UNEP], 2022); and Mexico was one of the countries that supported the proposal as a signatory to the Convention on Biological Diversity (CBD) (Secretaría de Relaciones Exteriores, 2022).

The territorial extension of protected areas within the country increased from 15.29 km^2 in 1917 to 980007.19 km^2 at the end of 2024, representing 19.2% of the total national territory (marine and terrestrial) (CONANP, 2025). Although 1986, 2023, and 2024 were the years when the most protected area decrees were issued, 2016 and 2017 were the years when the largest decrees were issued in terms of surface area, 508382.87 km^2 and 148087.80 km^2, respectively (Fig. 2.2) these 2 years concentrate 67.0% of the total area of protected areas decreed in the country. Considering the total area, the average area of protected areas decreed is 9074.14 km^2 per year, even though no protected areas were decreed in 56 years.

Protected areas in Mexico cover both land and marine areas. Over the years, between 1917 and 2024, of the total number of years in which natural protected areas were decreed (n = 52), 100% of the years (n = 52) included terrestrial areas in the decrees, and 36.5% of the years (n = 19) included marine areas. Decrees for protected areas that include terrestrial areas have been the most consistent over the years, averaging two decrees per year (Fig. 2.3a), while decrees for marine areas cover the largest surface areas (Fig. 2.3b). The increases in marine area averaged 6935.57 km^2 per year, despite fewer decrees, while the increase in terrestrial area averaged 2138.57 km^2 per year.

The Mexican territory is made up of 32 states or federal entities, and although by the end of 2024 there will be at least one federal protected area in each state, the evolution of decrees and territorial extension has varied over the years. Considering that the first protected area in Mexico was decreed in Mexico City in 1917, it was not until 1935 that three new protected areas were decreed: in the state of Mexico, in Morelos and in Puebla, one in each state.

In the first 20 years since the first protected area was established in 1917 and until 1936, most protected areas had been designated in the Estado de México (n = 4) and Mexico City (n = 4) (Fig. 2.4a). In the following 40 years, until the end of 1976, the cumulative number of protected natural areas was still concentrated in the state of Mexico (n = 10) and Mexico City (n = 9) in the two periods (1937–1956 and 1957–1976) (Fig. 2.4b and c).

Over the next 20 years, by the end of 1996, most protected areas had been designated in the state of Chiapas (n = 15) (Fig. 2.4d). By the end of 2016, most protected areas were located in Chiapas and Quintana Roo, both with 18 protected areas (Fig. 2.4e). Finally, by the end of 2024, protected areas are concentrated in Chiapas (n = 20) and Quintana Roo (n = 26) (Fig. 2.4f).

From the first protected area until the end of 1976, protected areas were concentrated in central Mexico, but after 1976 most decrees began to be distributed to other regions of the country. For example, between 1977 and 1996, most protected areas were not only decreed in more states in the central region, but also in states in the northwestern and southern regions of the country (Fig. 2.4d). However, this trend changed between 1997 and 2016, with fewer protected areas decreed in the central region and an increasing number of decrees in the northwestern and southern regions of the country, a trend that continued until 2024.

The territorial extension of protected areas has varied over the years. In 1917, there were 15.29 km^2 of protected areas, and by 2024, this had increased to 980007.19 km^2, of which 230965.63 km^2 are terrestrial. In the first 20 years since the first protected area was established in 1917 and until 1936, the territorial area designated as protected was 1724.05 km^2, mainly in the Estado de México (845.88 km^2) (Fig. 2.5a). In the following 40 years, until 1976, the cumulative territorial extension of protected areas was 57088.44 km^2, mostly in the states of Coahuila (16059.38 km^2), Chihuahua (7084.37 km^2), Nayarit (6872.16 km^2), Durango (6190.63 km^2), Zacatecas (5194.93 km^2), Jalisco (3404.44 km^2), and Sonora (2619.98 km^2) (Fig. 2.5b and c).

The increase in territorial extension of protected areas between 1917–1936 and 1937–1956 was 3157.4%, while between 1937–1956 and 1957–1976 it increased by only 1.7%. In the following 20 years, from 1977 to the end of 1996, most of the terrestrial area was in the states of Baja California (29844.64 km^2), Baja California Sur (25252.53 km^2), Coahuila (19804.92 km^2), Sonora (14030.66 km^2), Campeche (12858.76 km^2), and Chiapas (11265.91 km^2) (Fig. 2.5d), increasing by 193.0% compared to the total for the previous period.

By the end of 2016, the states with the largest territorial extension of protected areas were: Baja California (30233.14 km^2), Baja California Sur (25829.29 km^2), Coahuila (23284.68 km^2), Chihuahua (16223.5 km^2), Campeche (14222.97 km^2), Sonora (14030.66 km^2), and Chiapas (11401.84 km^2) (Fig. 2.5e), representing a 26.9% increase in protected terrestrial area compared to the previous period. In

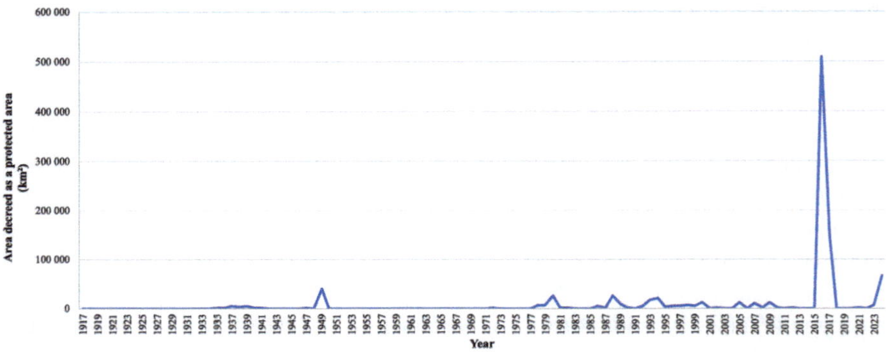

Fig. 2.2 Territorial extension of protected areas in Mexico, 1917–2024. **Source:** the authors with data from Comisión Nacional de Áreas Naturales Protegidas (2025)

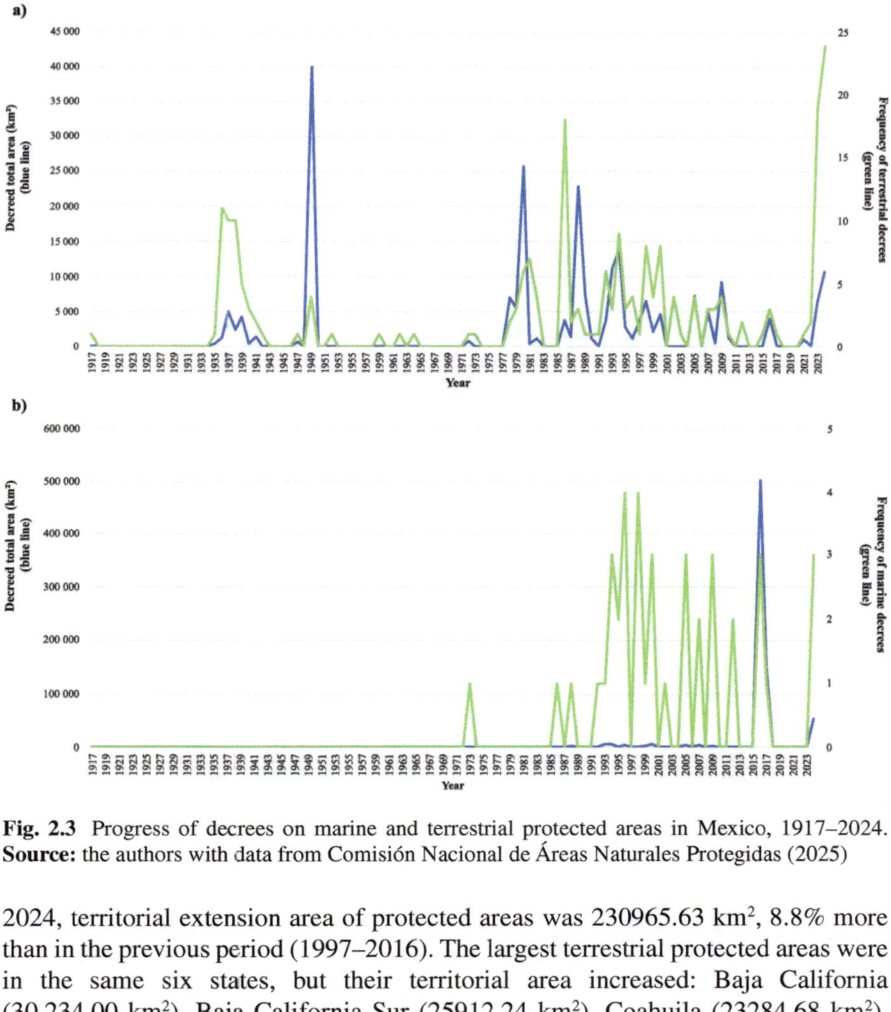

Fig. 2.3 Progress of decrees on marine and terrestrial protected areas in Mexico, 1917–2024. **Source:** the authors with data from Comisión Nacional de Áreas Naturales Protegidas (2025)

2024, territorial extension area of protected areas was 230965.63 km^2, 8.8% more than in the previous period (1997–2016). The largest terrestrial protected areas were in the same six states, but their territorial area increased: Baja California (30,234.00 km^2), Baja California Sur (25912.24 km^2), Coahuila (23284.68 km^2), Campeche (20015.91 km^2), Chihuahua (16227.69 km^2), Sonora (14030.66 km2), and Chiapas (11604.80 km2) (Fig. 2.5f).

In contrast to the number of protected area decrees, although until 1936 most of the territorial extension of protected areas was concentrated in the centre of the country, after 1937 most of the territorial extension of protected areas began to be distributed in other regions of the country. For example, from 1937 to 1976, most of the territorial extension was concentrated in states in the western and northwestern regions of Mexico (Fig. 2.5b and c). From 1977 to 2024, the largest amount of territorial extension of protected areas has been concentrated in the northern, northwestern, and southern regions of central Mexico, with an increasing number of decrees in the northwestern and southern regions of the country.

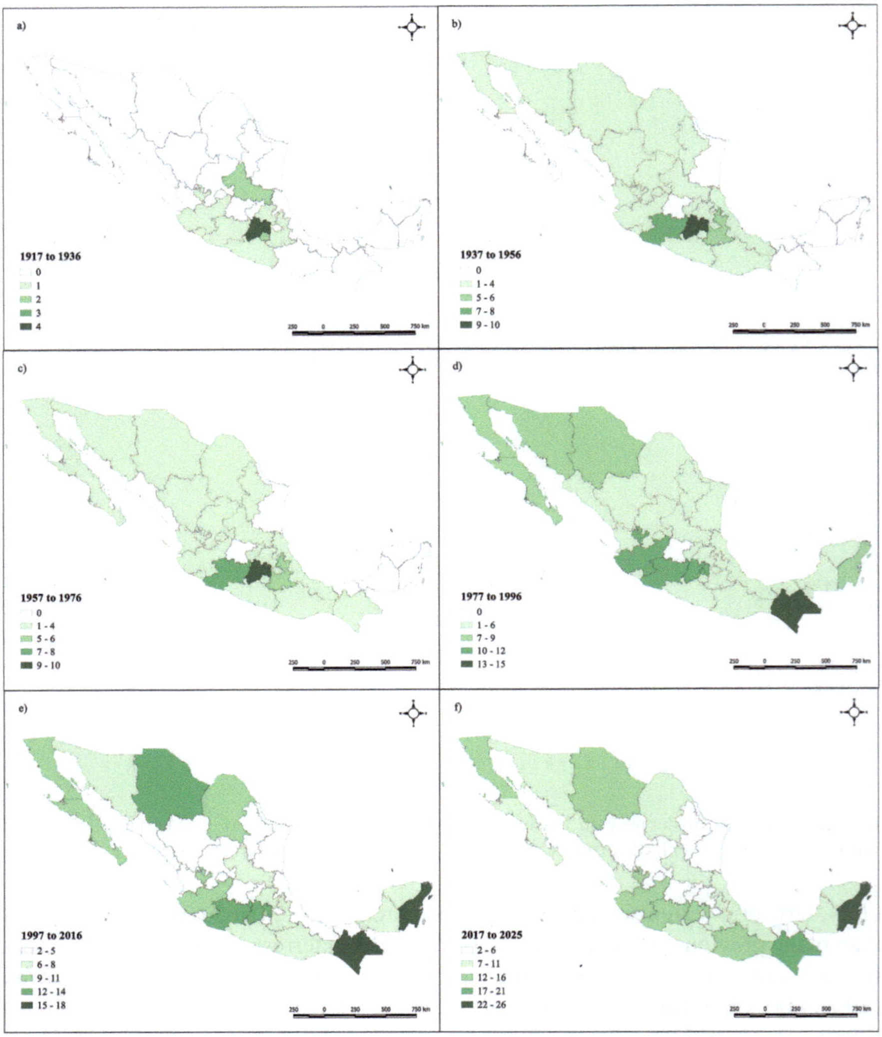

Fig. 2.4 Protected area decrees in Mexico by state in the period 1917–2024. **Source:** the authors with data from Comisión Nacional de Áreas Naturales Protegidas (2025)

The number of protected area decrees by category within the country has varied significantly over time (Fig. 2.6). Sanctuaries have the highest number of decrees in a single year, with 17 recorded in 1986. However, the categories of National Park ($n = 78$) and Flora and Fauna Protection Area ($n = 57$) are the categories that have been decreed the most since the first protected area in 1917 and until 2024.

On average, one decree per category was issued per year (with the exception of National Monuments). During the last 2 years in which protected areas were decreed (2023 and 2024), a similar trend was observed towards decreeing at least one type of category per year: 15 decrees for Flora and Fauna Protection Areas, 12 National Park decrees, 9 Sanctuary decrees, 6 Natural Resource Protection Area decrees, and 4 Biosphere Reserve decrees, all between 2023 and 2024 (Fig. 2.6), and National Monuments with only 5 decrees during the period analysed.

However, despite the historic increase in protected areas as a conservation instrument in Mexico, since 1984, the Mexican government has recognised that the state of conservation of protected areas in those years was deficient (50.0% of the protected areas decreed before 1984 showed serious deterioration due to the effects of

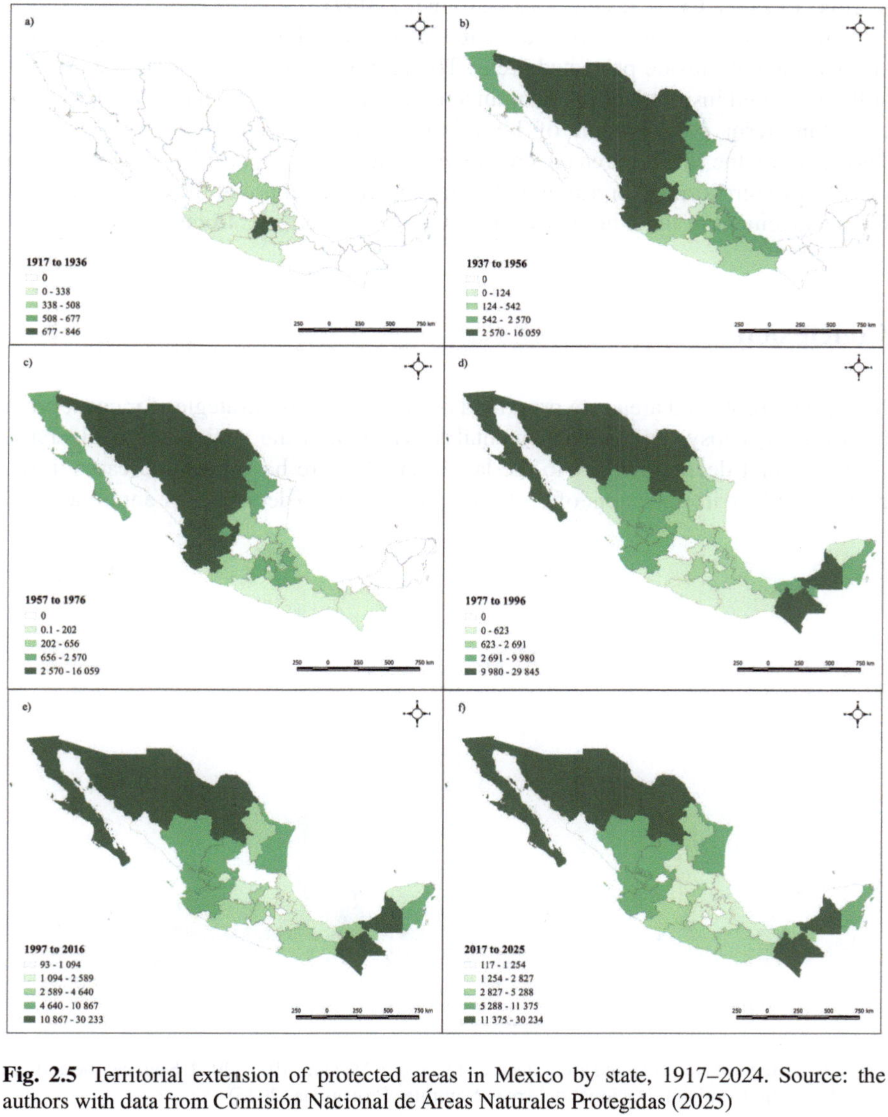

Fig. 2.5 Territorial extension of protected areas in Mexico by state, 1917–2024. Source: the authors with data from Comisión Nacional de Áreas Naturales Protegidas (2025)

logging, overgrazing, erosion, and poaching, among others) (Wyatt, 2015; CONANP, 2024; Carpio-Domínguez, 2024), and even the use of the land where the areas were located was legally irregular, and the capacity of personnel authorised to inspect protected areas was limited. Each inspector had to cover 61.84 km^2 without the minimum essential equipment, as only 20.0% of the areas had sufficient financial resources (Secretaría de Desarrollo Urbano y Ecología [SEDUE], 1984). It was not until the last decade of the twentieth century that the State began to consolidate its capacity to manage the protected areas adequately, starting to provide, among other things, sufficient personnel for management and planning instruments for legal protection (Bezaury-Creel & Gutiérrez Carbonell, 2009).

However, much remains to be done to safeguard biological and landscape diversity within and outside protected areas. The current coverage of protected areas is still considered insufficient to maintain a representative system that promotes effective, long-term conservation of both biodiversity and environmental services. Furthermore, the surface area covered by protected areas is considered to be only an indicator of territorial expression with legal coverage, but not of the effectiveness of the management of these areas (Bezaury-Creel y Gutiérrez Carbonell, 2009).

Conclusions

In Mexico, protected areas are presented as one of the best strategies for conserving biodiversity, ecosystems, environmental services, and the sociocultural diversity. From the first decree in 1917 to the last in 2024, there have been increases in the number and territorial extent of protected areas, with no decreases or annulments of

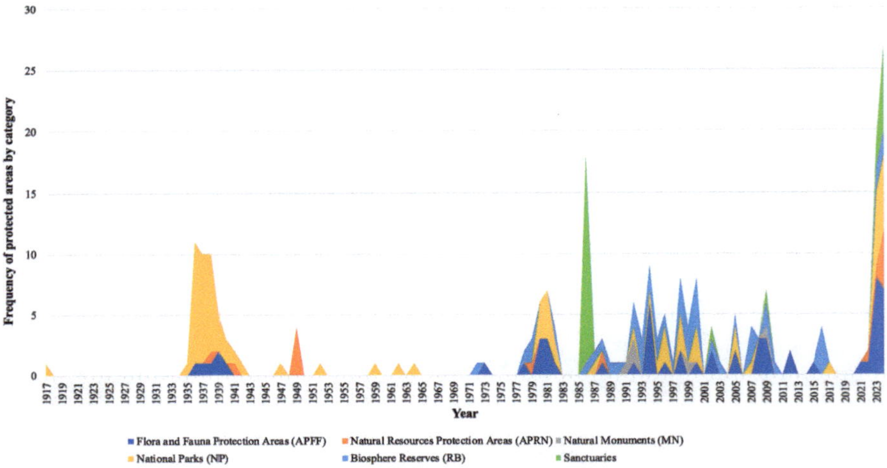

Fig. 2.6 Categories of protected areas decreed by year in Mexico, 1917–2024. **Source:** the authors with data from Comisión Nacional de Áreas Naturales Protegidas (2025)

decrees. In response to the changes caused by environmental degradation, particularly in rural areas, since the early 1990s, there has been a gradual intensification of federal environmental policy with protection and conservation criteria that have succeeded in consolidating more protected areas.

However, despite the historic progress made, there is still action to be taken and implemented to safeguard the biological and cultural diversity present in protected areas. Mexico is a megadiverse country in terms of ecosystems and species, and the increasing trend in the extension of protected areas appears to be insufficient to maintain a representative and complementary conservation and protection system in the long term.

Protected areas are essential, but they are not sufficient to contain environmental degradation and ensure biodiversity and sociocultural conservation. Increases in the number and extension of protected areas seem to be a good initiative but are not sufficient without a solid policy and administration of protected areas. The number and extension of protected areas are only an indicator of territorial expression, but do not represent the effectiveness of management in those areas.

References

Behrman, D. (1974). En asociación con la naturaleza; La Unesco y el medio ambiente. UNESCO. http://unesdoc.unesco.org/images/0013/001375/137549so.pdf

Bezaury-Creel, J. & Gutiérrez Carbonell, D. (2009). Áreas naturales protegidas y desarrollo social en México. In Comisión Nacional para el Conocimiento y Uso de la Biodiversidad (CONABIO), *Capital natural de México-Estado de conservación y tendencias de cambio* (pp. 385-431). CONABIO. Availbale on https://bioteca.biodiversidad.gob.mx/janium/Documentos/13334.pdf

Bishop, K., Dudley, N., Phillips, A., & Stolton, S. (2004). The uses and performance of the IUCN system of management categories for protected areas, IUCN: International Union for Conservation of nature.. Cardiff University, IUCN, Protected Areas Programme, UNEP-WCMC, WWF, Conservation International, Shell Corporation, British Petroleum (BP), International Council on Mining and Metals. https://portals.iucn.org/library/node/8520

Brenner, L. (2010). Gobernanza ambiental, actores sociales y conflictos en las Áreas Naturales Protegidas mexicanas. *Revista Mexicana de Sociología, 72*(2), 283–310. https://doi.org/10.22201/iis.01882503p.2010.002.17809

Carpio-Domínguez, J. (2021). Crimen Organizado (narcotráfico) y conservación ambiental: el tema pendiente de la seguridad pública en México. *Revista CS, 33*, 237–274. https://doi.org/10.18046/recs.i33.4076

Carpio-Domínguez, J. (2024). The harms and crimes of logging and deforestation. In *Oxford Research Encyclopedia of Criminology and Criminal Justice*. https://doi.org/10.1093/acrefore/9780190264079.013.769

Comisión Nacional de Áreas Naturales Protegidas (CONANP). (2012). *Biodiversidad Mexicana*. Available on http://www.biodiversidad.gob.mx

Comisión Nacional de Áreas Naturales Protegidas (CONANP). (2024). *Áreas Naturales Protegidas*. Available on https://www.gob.mx/conanp/documentos/areas-naturales-protegidas-278226

Comisión Nacional de Áreas Naturales Protegidas (CONANP). (2025). *Numeralia ANP*. Available on https://simec.conanp.gob.mx/numeralia.php

Comisión Nacional para el Conocimiento y Uso de la Biodiversidad (CONABIO). (2023a). *México megadiverso*. Available on https://www.biodiversidad.gob.mx/pais/quees

Comisión Nacional para el Conocimiento y Uso de la Biodiversidad (CONABIO). (2023b). *Áreas protegidas*. Available on https://www.biodiversidad.gob.mx/region/areasprot

Correa, A. (2016). Gobernanza de las áreas protegidas y su necesaria incorporación en la ley que crea el Servicio de Biodiversidad. *Revista de Derecho Ambiental, 6*, 224–239. https://revista-derechoambiental.uchile.cl/index.php/RDA/article/view/43324

Costa Pellizzaro, P., Antunes Hardt, L. P., Hardt, C., Hardt, M., & Assef Sehli, D. (2015). Stewardship and management of protected natural areas: The international context. *Journal Volume & Issue, 18*(1), 19–36. https://doi.org/10.1590/1809-4422ASOC509V1812015en

Dudley, N. (2008). *Guidelines for applying protected area management categories*. International Union for Conservation of nature (IUCN). https://portals.iucn.org/library/sites/library/files/documents/pag-021.pdf

Eagles, P., Romagosa, F., Buteau-Duitschaever, W. C., Havitz, M., Glover, T. D., & McCutcheon, B. (2012). Good governance in protected areas: An evaluation of stakeholders' perceptions in British Columbia and Ontario provincial parks. *Journal of Sustainable Tourism, 21*(1), 60–79. https://doi.org/10.1080/09669582.2012.671331

Ervin, J., Mulongoy, K. J., Lawrence, K., Game, E., Sheppard, D., Bridgewater, P., Bennett, G., Gidda, S. B., & Bos, P. (2010). Making protected areas relevant: A guide to integrating protected areas into wider landscapes, seascapes and sectoral plans and strategies. *Convention on Biological Diversity*. https://www.cbd.int/doc/publications/cbd-ts-44-en.pdf

Food and Agriculture Organization of the United Nations (FAO). (2014). *Protected areas, people and food security*. FAO. Available on https://openknowledge.fao.org/server/api/core/bitstreams/a83e1fee-532f-43d2-9133-ede6a5643888/content

Gobierno de México. (2013). *Plan Nacional de Desarrollo 2013–2018. Diario Oficial de la Federación*. Gobierno de México. Available on https://www.dof.gob.mx/nota_detalle.php?codigo=5299465&fecha=20/05/2013#gsc.tab=0

Gobierno de México. (2019). *Plan Nacional de Desarrollo 2019–2024*. Diario Oficial de la Federación. Gobierno de México. Available on https://www.dof.gob.mx/nota_detalle.php?codigo=5565599&fecha=12/07/2019#gsc.tab=0

Gravez, V., Heylings, P., Rivadeneira, C., & Zambrano, N. (2011). *Gobernanza en las Áreas Protegidas Marinas y Costeras: el caso del Ecuador*. Fundación Futuro Latinoamericano. Available on https://www.ffla.net/wp-content/uploads/2021/03/FFLA_2011_Gobernanza-en-Areas-Protegidas-Marinas-y-Costeras_Ecuador.pdf

Íñiguez Dávalos, L. I., Jiménez Sierra, C. L., Sosa Ramírez, J., & Ortega-Rubio, A. (2014). Categorías de las áreas naturales protegidas en México y una propuesta para la evaluación de su efectividad. *Investigación y Ciencia, 22*(60), 65–70. https://www.redalyc.org/pdf/674/67431160008.pdf

International Union for Conservation of Nature (IUCN). (2005). *Beneficios más allá de las fronteras. Actas del V Congreso Mundial de parques de la UICN*. Gland, Switzerland, and Cambridge, United Kingdom: IUCN. Available on https://redeuroparc.org/wp-content/uploads/2022/02/durban_actas_durban.pdf

International Union for Conservation of Nature (IUCN). (2025). *IUCN green list of protected and conserved areas*. IUCN. https://iucn.org/news/202311/iucn-director-general-appoints-iucn-green-list-panel

Kraus, D., Enns, A., Hebb, A., Murphy, S., Drake, D. A. R., & Bennett, B. (2023). Prioritizing nationally endemic species for conservation. *Conservation Science and Practice., 5*(1), 1–15. https://doi.org/10.1111/csp2.12845

Ley General del Equilibrio Ecológico y Protección al Ambiente (LGEEPA). (2024). *Última Reforma Publicada en el Diario Oficial de la Federación en abril 4 de 2024*. Congreso de la Unión. https://www.ordenjuridico.gob.mx/leyes.php#gsc.tab=0

Martínez, N., Espejel, I., & Martínez Valdés, C. (2016). Evaluation of governance in the Administration of Protected Areas on the peninsula of Baja California. *Frontera Norte, 28*(55), 103–129. https://www.scielo.org.mx/scielo.php?script=sci_arttext&pid=S0187-73722016000100005

Moya, F., Mella-Romero, J., & Simonetti, J. A. (2025). Do we have the necessary insights to classify species according to their conservation status? A case study in the world's southernmost lizard genus. *Journal for Nature Conservation, 83*. https://doi.org/10.1016/j.jnc.2024.126766

Murillo Flores, F. J., & Orozco Alvarado, J. (2006). *El turismo alternativo en las áreas naturales protegidas*. Universidad de Guadalajara. Available on http://www.cuc.udg.mx/sites/default/files/publicaciones/2006%20-%20El%20turismo%20alternativo%20en%20las%20%C3%A1reas%20naturales%20protegidas.pdf

Naughton-Treves, L., Buck Holland, M., & Brandon, K. (2005). The role of protected areas in conserving biodiversity and sustaining local livelihoods. *Annual Review of Environment and Resources, 30*, 219–252. https://doi.org/10.1146/annurev.energy.30.050504.164507

Noguera-Urbano, E. A. (2017). El endemismo: diferenciación del término, métodos y aplicaciones. *Acta Zoológica Mexicana, 33*(1), 89–107. https://doi.org/10.21829/azm.2017.3311016

Rivas Toledo, A. (2006). *Gobernanza de los Sistemas Nacionales de Áreas Protegidas en los Andes Tropicales: Diagnóstico regional y análisis comparativo*. IUCN. https://portals.iucn.org/library/sites/library/files/documents/2006-065.pdf

Sarukhán, P. K., Carabias, J., Rodolfo Dirzo, J. S., Llorente-Bousquets, J., Halffter, G., González, R., March, I., Mohar, A., Anta, S., & de la Maza, J. (2009). *Capital natural de México. Síntesis: conocimiento actual, evaluación y perspectivas de sustentabilidad*. Comisión Nacional para el Conocimiento y Uso de la Biodiversidad (CONABIO). Available on https://archive.org/details/CapitalNaturald00Saru/page/6/mode/2up

Secretaría de Desarrollo Urbano y Ecología (SEDUE). (1984). *Programa Nacional de Ecología*. Diario Oficial de la Federación. Available on https://www.dof.gob.mx/nota_detalle.php?codigo=4689824&fecha=26/09/1984&print=true

Secretaría de Medio Ambiente y Recursos Naturales (SEMARNAT). (2020). *Programa Sectorial de Medio Ambiente y Recursos Naturales 2020–2024*. Gobierno de México. Available on https://www.gob.mx/cms/uploads/attachment/file/566832/PROMARNAT-2020-2024.pdf

Secretaría de Relaciones Exteriores (SRE). (2022). *México apoya propuesta de la Presidencia de China e impulsa la adopción del Marco Mundial para la Biodiversidad en la COP15*. Comunicado No. 507. Gobierno de México. Available on https://www.gob.mx/sre/prensa/mexico-apoya-propuesta-de-la-presidencia-de-china-e-impulsa-la-adopcion-del-marco-mundial-para-la-biodiversidad-en-la-cop15?idiom=es

Secretariat of the Convention on Biological Diversity. (2008). *Protected areas in today's world: Their values and benefits for the welfare of the planet*. Secretariat of the Convention on Biological Diversity. Available on https://www.cbd.int/doc/publications/cbd-ts-36-en.pdf

The World Bank. (2010). Valuing protected areas. *The World Bank*. https://doi.org/10.13140/RG.2.1.2296.6642

United Nations. (2024). *Objetivos de Desarrollo Sostenible*. Available on https://www.un.org/sustainabledevelopment/es/objetivos-de-desarrollo-sostenible/

United Nations Environment Programme (UNEP). (2022). *Conferencia de las Partes del Convenio sobre la Diversidad Biológica (COP 15) (CBD/COP/15/L.25)*. United Nations. https://www.cbd.int/doc/c/2c37/244c/133052cdb1ff4d5556ffac94/cop-15-l-25-es.pdf

United Nations Environmental Programme-World Conservation Monitoring Centre (UNEP-WCMC). (2018). 2018 *United Nations list of protected areas*. Supplement on protected area management effectiveness. UNEP-WCMC. https://www.sprep.org/attachments/VirLib/Global/2018-list-protected-areas.pdf

Wyatt, T. (2015). Mapping the links between conflict and illegal logging. In A. Brisman, N. South, & R. White (Eds.), *Environmental crime and social conflict: Contemporary and emerging issues* (pp. 41–56). Routledge. https://doi.org/10.4324/9781315580012

Chapter 3
Geography of Natural Protected Areas

Introduction

Biological diversity, ecosystems, and natural resources are the main sustenance of human and non-human life; they are the natural capital in the Earth's bank, a capital whose loss would be irreversible. Protected areas provide protection and conservation not only for species and habitat diversity, but also preserve human cultural diversity. Protected areas in Mexico can be terrestrial or marine (including coasts) and are refuges for hundreds of species and ecosystems that could disappear if they were not within protected areas (Secretaría de Medio Ambiente y Recursos Naturales (SEMARNAT), 2011).

Protected areas in Mexico contribute to the conservation of natural heritage, but also to the knowledge, natural resources, and culture of indigenous peoples by promoting actions to reduce the impacts of deforestation, infrastructure construction, forest fires, land use change and the introduction of exotic or invasive species, among others (Comisión Nacional de Areas Naturales Protegidas [CONANP], 2017).

With the aim of describing the socio-environmental characteristics of current protected areas in Mexico, this chapter reviews and analyses federal protected areas by region from four perspectives: (1) categories and number of protected areas by region; (2) species and ecosystem biodiversity; (3) social and cultural diversity; and (4) political diversity.

Supplementary Information The online version contains supplementary material available at https://doi.org/10.1007/978-3-032-18218-0_3.

J. L. Carpio-Domínguez et al., *Drug Trafficking and Protected Areas in Mexico*,
https://doi.org/10.1007/978-3-032-18218-0_3

Protected Areas in Mexico: Surface area, Categories, and Territorial Distribution

Area of Protected Areas in Mexico

In 2024, there are 232 protected areas (terrestrial and marine) in Mexico (CONANP, 2025a). Of the total number of protected areas decreed, the largest percentage are exclusively terrestrial ($n = 192$, 82.8%), with a smaller number designated as marine protected areas ($n = 9$, 3.9%); while most protected areas are a combination of both (terrestrial and marine) ($n = 31$, 13.4%) (Fig. 3.1).

However, when analysing the territorial area decreed as protected areas, marine protected areas cover the largest territorial area: 749041.6 km^2, corresponding to 76.4% of the territorial area of protected areas decreed in Mexico, compared to the terrestrial area decreed, which represents 23.6% of the total (Table 3.1).

Protected areas have regulated and clearly defined territorial boundaries for biodiversity conservation, land use, and human occupation in accordance with local socio-environmental characteristics and management objectives defined to achieve the required level of environmental protection for each area (Milano, 2001; Medeiros, 2006).

Protected areas in Mexico are distributed throughout the national territory, with clearly defined boundaries that can be observed spatially, with marine protected areas covering the largest surface area. Some terrestrial protected areas exceed 2155.7 km^2, while marine protected areas can cover up to 436,141.2 km^2 (Fig. 3.2).

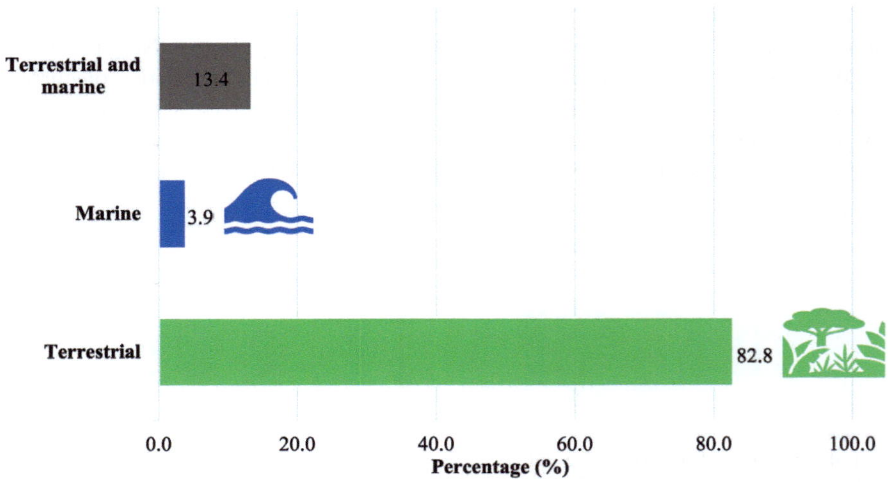

Fig. 3.1 Percentage of protected areas by type in Mexico until 2024. **Source:** the authors with data from Comisión Nacional de Áreas Naturales Protegidas (2025a)

Table 3.1 Area of national territory in protected areas in Mexico until 2024

Protected area	Area (km^2)	Percentage of total protected areas at national level	Percentage of total national territory
Terrestrial	230965.6	23.6	11.7%
Marine	749041.6	76.4	23.8%
Total	**980007.2**	**100**	

Source: the authors with data from Comisión Nacional de Áreas Naturales Protegidas (2025a)

Categories of Protected Areas

In Mexico, the 232 protected areas designated can be classified into six categories according to their use and sustainable management: 79 National Parks (PN), 57 Flora and Fauna Protection Areas (APFF), 48 Biosphere Reserves (RB), 28 Sanctuaries, 15 Natural Resource Protection Areas (APRN), and 5 Natural Monuments (MN) (Table 3.2), each with specific protection objectives.

Flora and Fauna Protection Areas (APFF) are the most decreed, but Biosphere Reserves (RB) (n = 48) cover the largest land area, 640337.1 km^2, representing 65.3% of the total area of protected areas in Mexico (see Table 3.2). Natural Monuments (MN), in contrast, are not only the least common type of protected area in Mexico (n = 5) and cover the smallest area (162.7 km^2), representing 0.02% of the total area of protected areas in the country.

When planning to decree a natural space as a protected area, one of the first steps is to determine the legal regime governing the land or marine space to be assessed or declared (Lausche, 2012). In Mexico, protected area categories are divisions established to regulate activities and restrictions in each area, and the objectives, types of activities, and management practices that apply to each category of protected area are established in general terms.

Each of the categories is subject to special protection, conservation, restoration, and development regimes, which restrict access and use of flora, fauna, and other natural resources in protected areas in order to ensure the ecological equilibrium and continuity of evolutionary and ecological processes (SEMARNAT, 2021). In addition to the general categories, the management programmes for each protected area specify restrictions and management actions specific to each area (Table 3.3).

Most of the territorial extension of protected areas belongs to the Biosphere Reserve category, which also covers mostly marine rather than terrestrial areas (Fig. 3.3). The other category that mostly covers marine areas is National Monuments.

Geographical Distribution of Protected Areas in Mexico by State

The 32 federal entities that constitute Mexican territory have at least one category of protected area until 2024. Baja California is the entity with the largest territorial extension assigned to these areas, with 30,234.0 km^2 of protected area, while Mexico City is the entity with the smallest territorial extension (341.1 km^2) (Fig. 3.4).

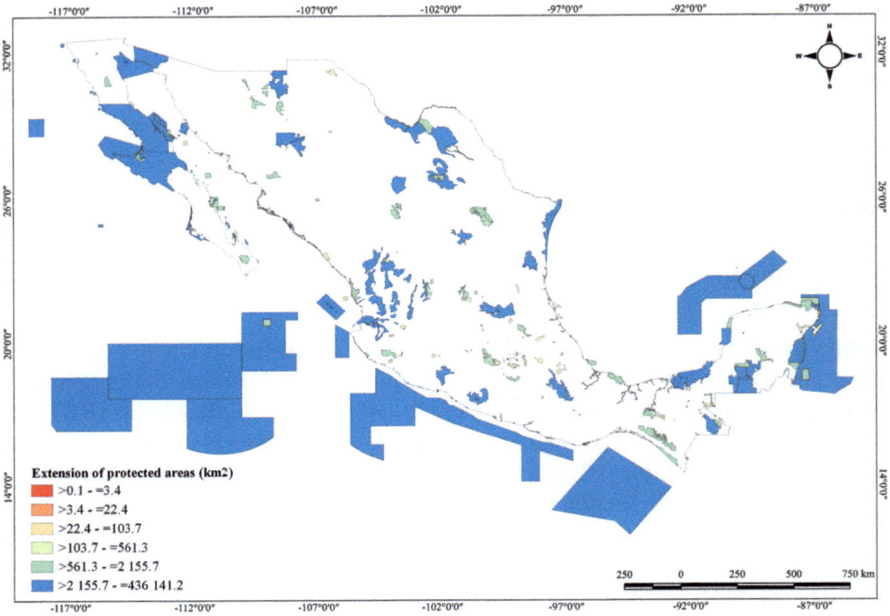

Fig. 3.2 Geographical distribution of protected areas in Mexico. **Source:** the authors with data from Comisión Nacional de Áreas Naturales Protegidas (2025a)

Table 3.2 Federal protected areas by category in Mexico until 2024

Category	Total of protected areas	Area (km^2)	Percentage of total protected territory	Percentage of the national total
Flora and Fauna Protection Areas (APFF)	57	75305.5	7.7	1.5
Natural Resources Protection Areas (APRN)	15	45852.9	4.7	0.9
Natural Monuments (MN)	5	162.7	0.02	0.0
National Parks (PN)	79	216794.1	22.1	4.2
Biosphere Reserves (RB)	48	640337.1	65.3	12.5
Sanctuaries	28	1554.9	0.16	0.0
Total	**232**	**980007.2**	**100**	**19.2**

Source: the authors with data from Comisión Nacional de Áreas Naturales Protegidas (2025a)

Considering the territorial distribution, Baja California was the state with the highest percentage of protected territorial extension, 13.1% of the total territory of protected areas, while Mexico City is the state with the smallest territorial area, with

Table 3.3 General restrictions by category of protected areas in Mexico

	Main restrictions
Flora and Fauna Protection Areas (APFF)	In these protected areas, only activities related to the preservation, repopulation, propagation, acclimatisation, shelter, research, and sustainable use of species, as well as those related to environmental education, can be implemented. Communities can use natural resources within these areas in accordance with Official Mexican Standards (Normas Oficiales Mexicanas, NOM) and land uses established in the protected area decree.
Natural Resources Protection Areas (APRN)	Only activities related to the preservation, protection, and sustainable use of natural resources, scientific research, recreation, tourism, and ecological education may be conducted in these protected areas.
Natural Monuments (MN)	Only activities related to conservation, scientific research, leisure, and education are permitted in these protected areas.
National Parks (PN)	In these protected areas, only activities related to the protection of natural resources, the enhancement of flora and fauna, and, in general, the preservation of ecosystems and their elements, as well as scientific research, recreation, tourism, and ecological education can be implemented.
Biosphere Reserves (RB)	These protected areas contain core zones where only activities related to the preservation of ecosystems and their elements and environmental education are authorised. Any activities that alter the ecosystems are prohibited. They contain buffer zones where only productive activities undertaken by the communities living there at the time the Area was declared can be conducted.
Sanctuaries (SANT)	In these protected areas, only scientific research, leisure, and environmental education activities that are compatible with the nature and characteristics of the area are permitted.

Source: the authors with data from Ley General de Equilibrio Ecológico y Protección al Ambiente (LGEEPA, 2024) (articles from 48 to 61) and the CONANP (2024)

0.05% of the total (Fig. 3.5). The states with the largest and smallest areas of protected land are located in the north and south of the country, respectively (Fig. 3.5).

The Southeast region has the highest number of protected areas (n = 91), followed by the Central region (n = 51), West (n = 51), Northwest (n = 51), and Northeast (n = 28) (Fig. 3.6). In the Southeast, Quintana Roo (n = 26) and Chiapas (n = 20) are the states with the most protected areas, and in the region the majority are Flora and Fauna Protection Areas (n = 28), followed by Biosphere Reserves (n = 23) and National Parks (n = 23).

In the Central region, the state of Mexico (n = 15) and Mexico City (n = 11) are the states with the most protected areas, and in the region, mostly National Parks (n = 34) and Biosphere Reserves (n = 8). In the West, Jalisco (n = 12), Michoacán (n = 12), and Nayarit (n = 7) are the states with the most protected areas, and the region is mostly Biosphere Reserves (n = 13), National Parks (n = 13), and Sanctuaries (n = 9).

In the Northwest, Chihuahua (n = 14) and Baja California (n = 12) are the states with the most PNAs, and the region mostly has Flora and Fauna Protection Areas

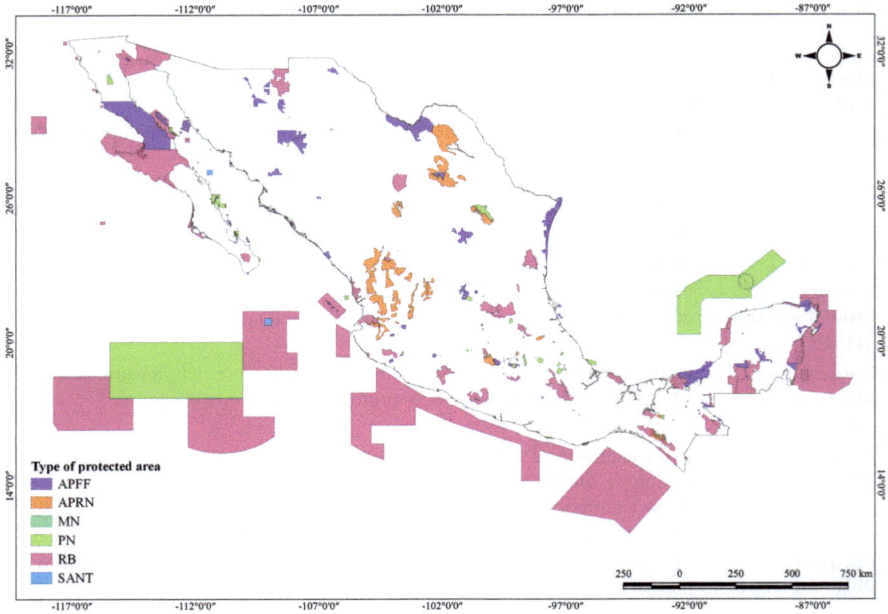

Fig. 3.3 Geographical distribution of protected areas by category until 2024. **Source:** the authors with data from Comisión Nacional de Áreas Naturales Protegidas (2025a). Abbreviations: Flora and Fauna Protection Areas (APFF); Natural Resources Protection Areas (APRN); Natural Monuments (MN); National Parks (PN); Biosphere Reserves (RB); Sanctuaries (SANT)

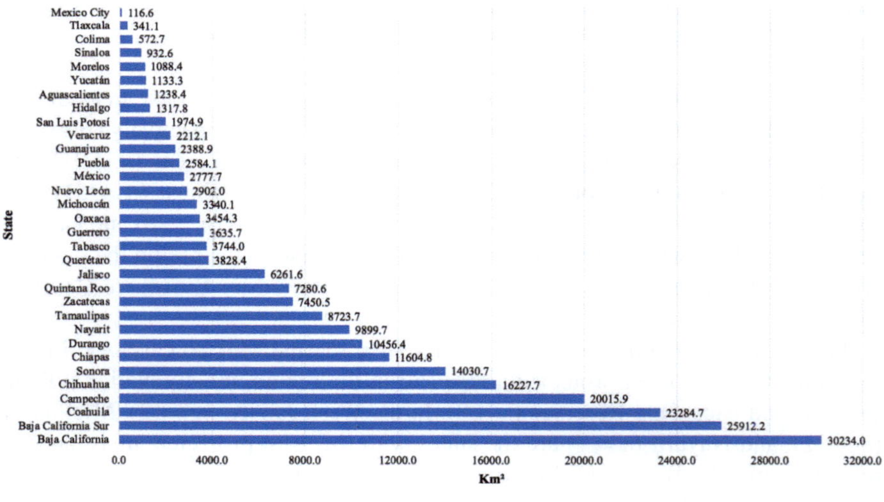

Fig. 3.4 Territorial extension of protected areas by state until 2024. **Source:** the authors with data from Comisión Nacional de Áreas Naturales Protegidas (2025a)

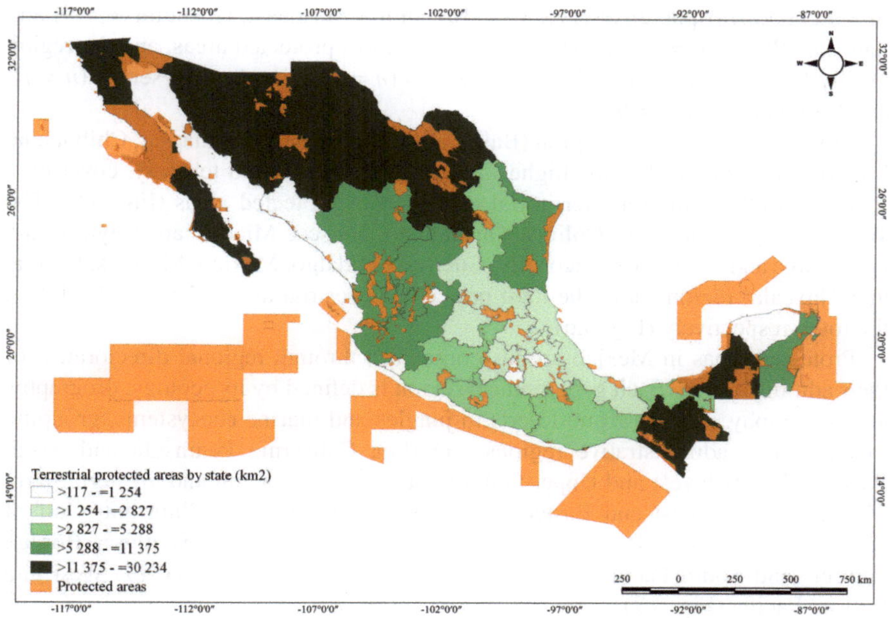

Fig. 3.5 Geographical location of protected areas by state in Mexico until 2024. **Source:** the authors with data from Comisión Nacional de Áreas Naturales Protegidas (2025a)

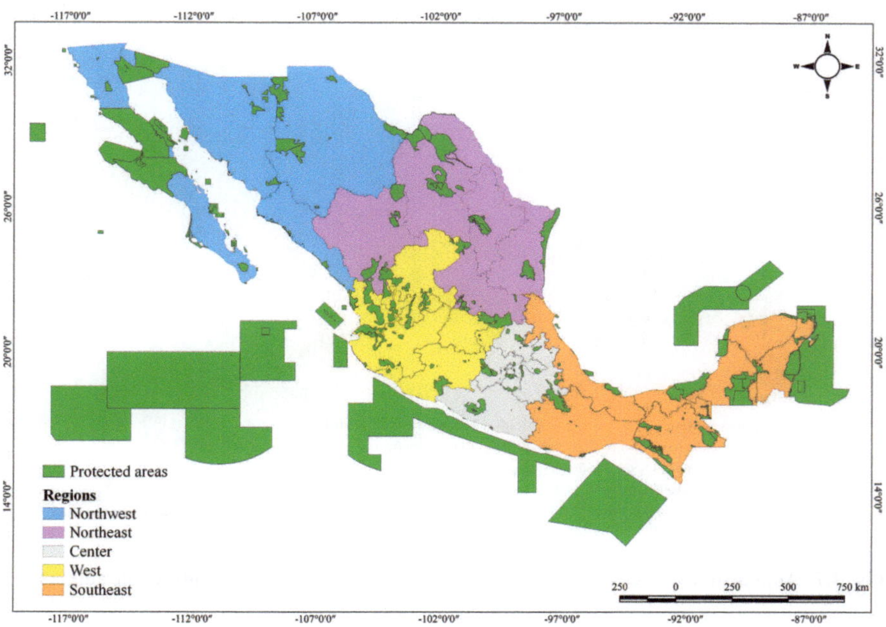

Fig. 3.6 Regionalisation of protected areas by state in Mexico until 2024. **Source:** the authors with data from Comisión Nacional de Áreas Naturales Protegidas (2025a)

(n = 22) and Biosphere Reserves (n = 15). In the Northeast, Coahuila (n = 9) and San Luis Potosí (n = 7) are the states with the most protected areas, and the region mostly has Flora and Fauna Protection Areas (n = 7), Biosphere Reserves (n = 7), and National Parks (n = 6).

However, the Northeast region (Baja California, Baja California Sur, Chihuahua, Sinaloa, and Sonora) has the highest percentage of protected territory, covering a total of 37.8% of the total territorial extension of protected areas (Fig. 3.6). The Western (Aguascalientes, Colima, Guanajuato, Jalisco, Michoacán, Nayarit, and Querétaro) and Central (Mexico City, Guerrero, Hidalgo, Mexico, Morelos, Puebla, and Tlaxcala) regions have the least protected terrestrial area, 15.1% and 5.1% of the total, respectively (Fig. 3.6).

Protected areas in Mexico are also organised through regional directorates for their administration. Each geographical region is defined by its ecology, orography, and geography, ranging from deserts to jungles and marine ecosystems, grouping together nine administrative regions: (1) Baja California Peninsula and North Pacific, (2) Northwest and Upper Gulf of California, (3) North and Western Sierra Madre, (4) Northeast and Eastern Sierra Madre, (5) Coastal Plain and Gulf of Mexico, (6) Yucatan Peninsula and Mexican Caribbean, (7) Southern Border, Isthmus and South Pacific, (8) Centre and Neovolcanic Axis, and (9) West and Central Pacific (Fig. 3.7).

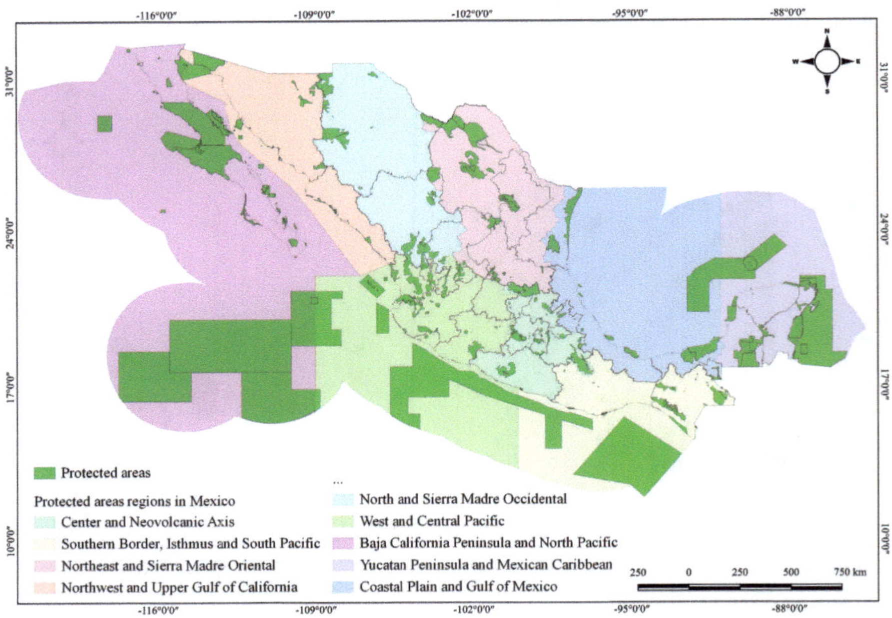

Fig. 3.7 Administrative regions of protected areas in Mexico until 2024. **Source:** the authors

(Bio) Diversity in Mexico's Protected Areas

Of the 232 protected areas in Mexico, 26.7% (n = 62) have species designated as microendemic,[1] 25.2% (n = 121) report having endemic species, and 84.1% (n = 195) indicate having representative species of flora and/or fauna (see **Supplementary Material 3.1**). In addition to being internationally recognised by the IUCN, some protected areas have international designations and/or certificates. 23.7% (n = 55) of protected areas have at least one international designation or certificate (16.4% are recognised as RAMSAR-Wetlands Convention sites, 1.3% have World Heritage World Heritage status, 5.6% are MaB—Man and Biosphere sites, and 1.0% are on the Green List), 14.7% (n = 34) have two designations or certificates, and 3.0% (n = 7) have three international designations or certificates (see **Supplementary Material 3.1**).

In protected areas, it is possible to identify various types of vegetation or ecosystems. The main type of vegetation ecosystem present in Mexico's protected areas is the territorial sea, which covers 72.2% (707944.2 km^2) of the total area of protected areas, while the least common vegetation or ecosystem is the mesophilic mountain forest (0.3%, 2846.7 km^2). (Table 3.4).

Table 3.4 Main types of vegetation, land use, and ecosystems in protected areas in Mexico until 2024

Vegetation type or ecosystem	Surface (km^2)	Percentage protected by protected area (%)
Mesophilic Mountain Forest	2846.7	0.3
Mangrove	4722.7	0.5
Grassland	6769.8	0.7
Hydrophilic vegetation	9949.7	1.0
No apparent vegetation	11314.6	1.2
Agroecosystems	15035.9	1.5
Other types of vegetation	24397.9	2.5
Deciduous Forest	25416.5	2.6
Coniferous Forest	30222.0	3.1
Oak Forest	32282.4	3.3
Xerophytic scrubland	80070.4	8.2
Territorial Sea	707944.2	72.2
Total	**980007.2**	**100**

Source: the authors with data from Comisión Nacional de Áreas Naturales Protegidas (2025a)

[1] Microendemic species have a very limited distribution, are established in a specific zone, in a very small geographical area, and are not found elsewhere.

Protected Areas with Special Characteristics

Although all protected areas in Mexico are important for achieving the objectives of conservation and preservation of nature, ecosystems, and diverse forms of life, as well as the culture of society, the CONANP (2025a) highlights the following 21 protected areas among all others for their most special characteristics:

(i) Desierto de los Leones (National Park), Mexico's first protected area.
(ii) Revillagigedo (National Park) provides protection for the largest marine area.
(iii) Valle de los Cirios (Protected Natural Area) is the largest terrestrial area.
(iv) Pico de Orizaba (National Park) provides protection at the highest altitude.
(v) Sierra Gorda -Querétaro- (RB) provides protection for the largest number of vegetation types.
(vi) Banco Chinchorro (RB) provides protection for the largest area of coral reefs.
(vii) Janos Brinda (RB) provides protection for the largest area of grasslands.
(viii) Calakmul (RB) provides protection for the largest area of tropical forests.
(ix) Laguna de Términos (APFF) provides protection for the largest area of mangroves.
(x) El Triunfo (RB) provides protection for the largest area of mesophilic mountain forests.
(xi) Tehuacán-Cuicatlán (RB) provides protection for the greatest diversity of cacti and agaves.
(xii) Gulf of California Islands (APFF) provides protection for the greatest diversity of seabirds.
(xiii) Palenque (PN) provides protection for the greatest diversity of birds of prey.
(xiv) Tuxtlas (RB) provides protection for the greatest diversity of ants.
(xv) Montes Azules (RB) provides protection for the greatest diversity of mammals and butterflies.
(xvi) Selva El Ocote (RB) provides protection for the greatest diversity of hummingbirds.
(xvii) Marismas Nacionales (RB) provides protection for the greatest diversity of Anseriformes (geese and ducks).
(xviii) Ojo de Liebre Lagoon Complex (RB) provides protection for the main breeding area of the grey whale (*Eschrichtius robustus*).
(xix) Rancho Nuevo Beach (SANT) provides protection for the only beach where the Kemp's ridley sea turtle (*Lepidochelys kempii*) nests.
(xx) Sierra San Pedro Mártir (PN) provides protection for the main reintroduction area for the California condor (*Gymnogyps californianus*).
(xxi) Mariposa Monarca (RB) provides protection for the main winter refuge sites for the monarch butterfly (*Danaus plexippus*).

Human Communities in Mexico's Protected Areas

Areas decreed as protected areas by authorities are inhabited or have human populations in their vicinity. In countries with indigenous populations, protected areas are often located in territories that have been used and cared ancestrally by the indigenous peoples settled there, which can lead to conflicts and must be considered for their management (Schmidt, 2010). Protected areas are essential for maintaining biodiversity, but they must also incorporate the concerns of the local communities that inhabit them.

Biodiversity and ecosystem conservation objectives must take social objectives into account, especially those of communities. For the management of protected areas, it is essential to understand not only the ecosystemic dimension of the area but also its social and cultural dimensions and the various interests and needs of the population that depends directly or indirectly on these resources (Gravez et al., 2011). Protected areas cover large territorial extensions, where the legitimate presence of rural and/or indigenous communities has been common for many years under the command of local governments and the influence of communities and cities of various levels of government (Rivas Toledo, 2006).

It is common for protected areas to be home to rural and/or indigenous populations that have close economic and sociocultural ties to biodiversity (Naughton-Treves et al., 2005; Martínez et al., 2016) (Fig. 3.8). The relationship between nature and culture over thousands of years has included the use and exploitation of natural territories, transforming them into cultural landscapes, which in turn has given rise to beliefs, rituals, spiritual practices, production and consumption patterns related to life cycles and differentiated in each territory of the planet (Vásquez Sánchez, 2017).

Resources in protected areas are often also essential for ensuring healthy communities and are even the main source of economic income for many communities, as millions of people depend on protected areas for their livelihoods. Some benefit directly, through the consumption of food produced or obtained in or around protected areas, and others indirectly, through employment and income, which provide benefits that contribute to maintaining livelihoods (Secretariat of the Convention on Biological Diversity, 2008; The World Bank, 2010; FAO, 2014).

Within protected areas there are rural and urban population areas. In protected areas located in the west, centre, and south of the country, there are more population areas within protected areas. The urban population is mostly located on the periphery of protected areas, but there are states with at least one urban area within protected areas (Fig. 3.9 and Table 3.5).

All states have at least one protected area, and there is at least one protected area within a municipality for each state. Baja California Sur and Quintana Roo are the states with the highest number of protected areas within their municipalities; in both states, all municipalities report having at least one protected area. Nayarit and Querétaro are the states with the most communities within protected areas, 11.4% and 20.9%, respectively, of the total number of communities per state. In the states of Aguascalientes, Baja California, Colima, Durango, Guerrero, Quintana Roo,

Fig. 3.8 Raramuri person in the Chihuahua mountains selling handicrafts made from natural products and materials from the region. Photograph: José Luis Carpio Domínguez

Sinaloa, and Tabasco, all communities within protected areas are in rural communities (Table 3.5).

Mexico is home to 70 indigenous peoples, who speak more than 60 languages other than Spanish and live in all the country's federal entities. The Constitución Política de los Estados Unidos Mexicanos (2025) (Political Constitution of the United Mexican States) recognises that Mexico has a multicultural and multi-ethnic composition originally based on its indigenous peoples, collectivities with a historical continuity from the pre-colonial societies established in the national territory, which preserve, develop, and transmit their social, regulatory, economic, cultural, and political institutions, or part of them.

Most indigenous peoples live in biodiverse territories. Approximately 70.0% of indigenous territories coincide with areas important for conservation, of which 26.3% correspond to protected areas, where 44 indigenous peoples live (CONANP, 2020). The indigenous peoples with the largest populations in protected areas in Mexico are the Nahuas, Mayas, Zapotecs, Mixtecs, Otomíes, Totonacas, Tsotsiles, Tzeltales, Mazahuas, Mazatecos, among others (Fig. 3.10).

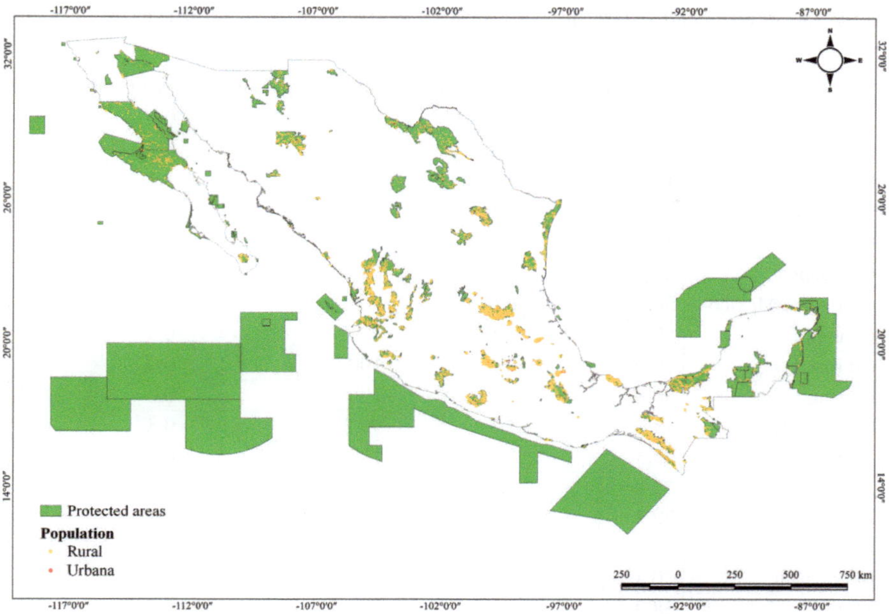

Fig. 3.9 Rural and urban areas in protected areas in Mexico until 2024. **Source:** the authors with data from Comisión Nacional de Áreas Naturales Protegidas (2025a) and Instituto Nacional de Estadística y Geografía (INEGI) (2025)

In 23 of Mexico's 32 states, at least one indigenous language is present within protected areas, with the states of Puebla ($n = 5$), Chiapas ($n = 8$), and Oaxaca ($n = 8$) having the greatest diversity of indigenous languages. However, the states of Nayarit ($n = 212$), Chiapas ($n = 199$), and Durango ($n = 158$) have the highest number of localities where at least one indigenous language within protected areas (Table 3.6). (Table 3.6).

Protected areas in the states of north-eastern Mexico account for the largest number of municipalities in the region; 40.9% of all municipalities in the northeast have at least one protected area. Of all the localities located in the central region of the country, 4.3% are located in at least one protected natural area, and it is in the protected areas of the southeastern region where the greatest diversity of indigenous languages exists (Table 3.7).

Management models for protected areas must understand the needs, priorities, practices, and values of local and indigenous communities regarding the management objective or category of the protected area, establishing strategies and policies that equitably distribute the costs and benefits of conservation and sustainable use (Valverde Garnica, 2015).

Table 3.5 Rural and urban areas in protected areas in Mexico by state until 2024

State	% con ANP			
	Municipalities	Communities	Rural communities	Urban communities
Aguascalientes	54.4	0.8	100	0.0
Baja California	71.4	2.2	100	0.0
Baja California Sur	100	5.2	98.6	1.4
Campeche	84.6	5.8	97.6	2.4
Chiapas	33.1	4.9	99.6	0.4
Chihuahua	29.9	1.5	99.4	0.6
Ciudad de México	56.3	0.6	90.9	9.1
Coahuila	50.0	2.8	99.7	0.3
Colima	70.0	0.4	100	0.0
Durango	30.8	3.2	100	0.0
Guanajuato	13.0	2.3	99.2	0.8
Guerrero	23.5	2.6	100	0.0
Hidalgo	25.0	4.4	97.9	2.1
Jalisco	39.2	1.9	99.5	0.5
México	40.8	6.9	95.2	4.8
Michoacán	20.4	1.5	98.2	1.2
Morelos	38.9	10.5	96.4	3.6
Nayarit	95.0	11.4	99.8	0.2
Nuevo León	23.5	1.9	99.5	0.5
Oaxaca	9.6	1.2	96.1	3.9
Puebla	21.7	3.9	96.4	3.6
Querétaro	66.7	20.9	99.0	1.0
Quintana Roo	100	1.3	100	0.0
San Luis Potosí	33.9	0.9	97.6	2.4
Sinaloa	55.0	0.3	100	0.0
Sonora	40.3	0.4	98.6	1.4
Tabasco	35.3	6.3	100	0.0
Tamaulipas	20.9	1.9	99.6	0.4
Tlaxcala	23.3	3.5	91.8	8.2
Veracruz	18.4	1.9	97.0	3.0
Yucatán	11.3	0.2	85.7	14.3
Zacatecas	51.7	2.5	99.3	0.7

Source: the authors with data from INEGI (2025)

Management Programmes in Protected Areas in Mexico

The designation of a protected area is insufficient to ensure conservation; complementary policy and management instruments are necessary once the protected area has been decreed (Dudley, 2008; Martínez et al., 2016). The instruments that determine conservation and use strategies for protected areas worldwide are management plans or programmes, conservation programmes, conservation and management

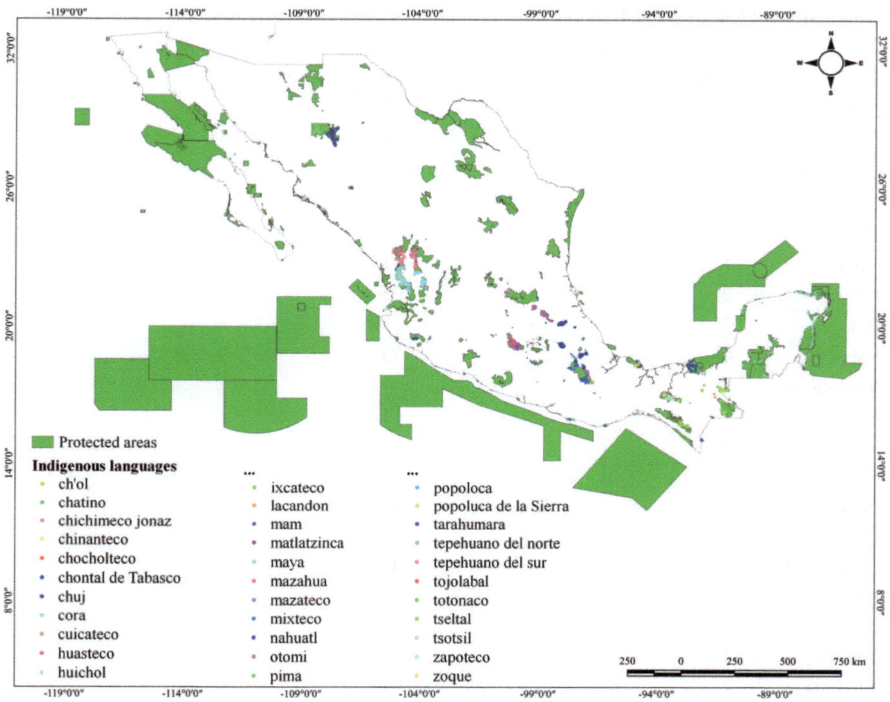

Fig. 3.10 Indigenous languages and protected areas in Mexico. Source: the authors with data from Comisión Nacional de Áreas Naturales Protegidas (2025a)

programmes, master plans, guiding plans, etc. In Mexico, these instruments are known as management programmes (CONANP, 2025b). Management programmes are the guiding public policy instrument for planning and regulation that establishes the basic activities, actions, and guidelines for the management and administration of the respective protected area. These programmes are crucial for ensuring the protection and conservation of ecosystems and biodiversity in these areas, as well as for guiding the productive activities permitted within protected areas under a context of prohibitions and conditions (CONANP, 2021).

In addition, the protected areas management programme aims to work with the social, public, and private sectors to promote and encourage surveillance and monitoring activities to preserve and protect natural resources and to carry out technical studies and biological monitoring activities in these areas (CONANP, 2021). In general, management programmes consist of three components: (1) Community Surveillance Component (CVC), (2) Biological Monitoring Component (CMB), and (3) Protected Area Strengthening Component (CFOR) (CONANP, 2021).

Article 65 of the Ley General del Equilibrio Ecológico y la Protección al Ambiente (LGEEPA) (2024) (General Law of Ecological Balance and Environmental Protection) stipulates that all protected areas under federal jurisdiction must have a

Table 3.6 Indigenous languages in localities with protected areas by state

State	Indigenous languages in protected areas	Localities with indigenous languages in protected areas
Campeche	1	17
Chiapas	8	199
Chihuahua	3	116
Durango	3	158
Guanajuato	1	5
Guerrero	1	1
Hidalgo	2	105
Jalisco	2	78
México	4	73
Michoacán	2	4
Morelos	1	1
Nayarit	4	212
Oaxaca	8	108
Puebla	5	147
Querétaro	1	1
Quintana Roo	1	47
San Luis Potosí	2	2
Sonora	1	15
Tabasco	3	61
Tlaxcala	1	20
Veracruz	2	103
Yucatán	1	33
Zacatecas	1	18

Source: the authors with data from Comisión Nacional de Áreas Naturales Protegidas (2025a)

Table 3.7 Municipalities, localities, and indigenous languages in protected areas by region in Mexico by state

Region	%		Indigenous languages in protected areas
	Municipalities in protected areas	Localities in protected areas	
Northwest	40.9	1.4	4
West	37.9	3.5	11
Centre	32.9	4.3	14
Northeast	31.9	2.1	5
Southeast	16.6	2.8	24

Source: the authors with data Comisión Nacional de Áreas Naturales Protegidas (2025a)

corresponding Management Programme, and according to Article 66 of the LGEEPA (2024), the content of management programmes must include:

I. Description of the physical, biological, social, and cultural characteristics of the protected area, in the national, regional, and local context.
II. The actions to be conducted in the short, medium, and long term include the following: environmental research and education; protection and sustainable use of natural resources, flora, and fauna; development of recreational and tourist activities, infrastructure works, and other productive activities; financing for the administration of the area; contingency prevention and control; surveillance; and any other actions required due to the characteristics of the protected area.
III. Organise the administration of the area and the mechanisms for participation of individuals and communities settled therein, as well as all those persons, institutions, groups, and social organisations interested in its protection and sustainable use.
IV. The specific objectives of the protected area.
V. Reference to the official Mexican standards (NOMs) applicable to each and every one of the activities to which the area is subject;
VI. Existing biological inventories and those planned and.
VII. The administrative rules governing activities conducted in the protected area.

The LGEEPA establishes that all protected areas in Mexico must have a management programme drafted by SEMARNAT within 1 year of the publication of the respective declaration in the Official Gazette of the Federation (Article 65 of the LGEEPA, 2024); however, although there are designated protected areas, not all of them have been provided with a management programme. Not all protected areas in Mexico have a management programme. Of the 232 protected areas designated, 54.7% ($n = 127$) have management programmes until 2024. Of the total number of National Monuments (MN), all have management programmes, while Sanctuaries (SANT) have the fewest management programmes (10.7%) (Fig. 3.11).

Conclusions

In Mexico, there are 232 protected areas until 2024. Most of the protected areas are located in marine zones; however, terrestrial protected areas cover a wide variety of vegetation types and ecosystems. In addition, many of the protected areas are home to important flora and fauna species, including endemic species. The establishment of protected areas is of paramount importance in the preservation of both ecosystems and species of flora and fauna. Furthermore, these areas are also of significant socio-environmental value.

Mexico's protected areas are sociocultural territories where rural and indigenous communities live, interact, and utilise ecosystems and natural species as a means of

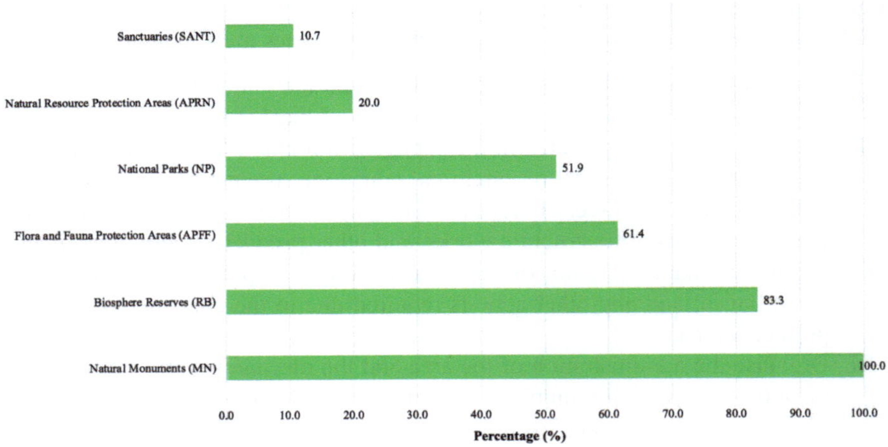

Fig. 3.11 Percentage of protected areas with management programmes in Mexico by category until 2024. **Source:** the authors with data from Comisión Nacional de Áreas Naturales Protegidas (2025a)

subsistence, becoming areas that also contribute to the preservation of traditional and ancestral culture.

However, the number of protected areas is not sufficient; policy and public administration instruments, such as management programmes, are also needed to help achieve the objectives. In Mexico, although the law establishes that protected areas must have a management programme drafted by the relevant federal authorities, not all federal protected areas in the country have their respective management programmes.

Management programmes for protected areas are crucial for the conservation of biodiversity and the use of natural resources within the areas, but the management programmes also help to establish the strategies and actions necessary to achieve the objectives of each protected area, including activities for the use of resources by society and actions for the surveillance of the areas.

References

Comisión Nacional de Áreas Naturales Protegidas (CONANP). (2017). *Las Áreas Naturales Protegidas: Respuestas Naturales frente al Cambio Climático*. Gobierno de México. Available on https://www.gob.mx/conanp/documentos/las-areas-naturales-protegidas-respuestas-naturales-frente-al-cambio-climatico?state=published

Comisión Nacional de Áreas Naturales Protegidas (CONANP). (2020). *Los pueblos indígenas en las Áreas Naturales Protegidas*. Gobierno de México. Available on https://www.gob.mx/conanp/articulos/los-pueblos-indigenas-en-las-areas-naturales-protegidas?idiom=es

Comisión Nacional de Áreas Naturales Protegidas (CONANP). (2021). *Programa de Manejo de Áreas Naturales Protegidas (PROMANP)*. Gobierno de México. Available on https://

www.gob.mx/conanp/acciones-y-programas/programa-para-la-integracion-o-modificacion--de-los-programas-de-manejo-de-las-areas-naturales-protegidas-competencia-de-la-federacion-promanp

Comisión Nacional de Áreas Naturales Protegidas (CONANP). (2024). *Áreas Naturales Protegidas*. Gobierno de México. Available on https://www.gob.mx/conanp/documentos/areas-naturales-protegidas-278226

Comisión Nacional de Áreas Naturales Protegidas (CONANP). (2025a). *Sistema de Información, Monitoreo y Evaluación para la Conservación (SIMEC)*. Gobierno de México. Available on https://simec.conanp.gob.mx/

Comisión Nacional de Áreas Naturales Protegidas (CONANP). (2025b). *Programas de Manejo de las Áreas Naturales Protegidas de México*. Gobierno de México. Available on https://www.gob.mx/conanp/acciones-y-programas/programas-de-manejo

Constitución Política de los Estados Unidos Mexicanos (2025, 14 de abril). *Congreso de la Unión*. Diario Oficial de la Federación. https://www.ordenjuridico.gob.mx/leyes.php#gsc.tab=0

Dudley, N. (2008). *Guidelines for applying protected area management categories*. International Union for Conservation of Nature (IUCN). https://portals.iucn.org/library/sites/library/files/documents/pag-021.pdf

Food and Agriculture Organization of the United Nations (FAO). (2014). *Protected Areas, People and Food Security*. FAO. Available on https://openknowledge.fao.org/server/api/core/bitstreams/a83e1fee-532f-43d2-9133-ede6a5643888/content

Gravez, V., Heylings, P., Rivadeneira, C., & Zambrano, N. (2011). Gobernanza en las Áreas Protegidas Marinas y Costeras: el caso del Ecuador. *Fundación Futuro Latinoamericano.*. https://www.ffla.net/wp-content/uploads/2021/03/FFLA_2011_Gobernanza-en-Areas-Protegidas-Marinas-y-Costeras_Ecuador.pdf

Instituto Nacional de Estadística y Geografía (INEGI). (2025). *Censo de Población y Vivienda 2020*. Gobierno de México. Available on https://www.inegi.org.mx/

Lausche, B. (2012). *Directrices para la legislación relativa a áreas protegidas*. International Union for Conservation of Nature (UICN). https://portals.iucn.org/library/sitEs/library/filEs/documents/EPLP-081-Es.pdf

Ley General del Equilibrio Ecológico y Protección al Ambiente (LGEEPA). (2024). *Última Reforma Publicada en el Diario Oficial de la Federación el 1 de abril de 2024*. Honorable Cámara de Diputados del Honorable Congreso de la Unión. Gobierno de México. Available on https://www.diputados.gob.mx/LeyesBiblio/pdf/LGEEPA.pdf.

Martínez, N., Espejel, I., & Martínez Valdés, C. (2016). Evaluation of Governance in the Administration of Protected Areas on the Peninsula of Baja California. *Frontera Norte, 28*(55), 103–129. https://doi.org/10.17428/rfn.v28i55.179

Medeiros, R. (2006). Evolução das tipologias e categorias de áreas protegidas no Brasil. *Ambiente & Sociedade, 9*(1), 41–64. https://doi.org/10.1590/S1414-753X2006000100003

Milano, M. (2001). Parques e reservas: uma análise da política brasileira de unidades de conservação. *Floresta, 15*(1/2), 04–09. https://doi.org/10.5380/rf.v15i12.6353

Naughton-Treves, L., Buck Holland, M., & Brandon, K. (2005). The Role of Protected Areas in Conserving Biodiversity and Sustaining Local Livelihoods. *Annual Review of Environment and Resources, 30*, 219–252. https://doi.org/10.1146/annurev.energy.30.050504.164507

Rivas Toledo, A. (2006). *Gobernanza de los Sistemas Nacionales de Áreas Protegidas en los Andes Tropicales: Diagnóstico regional y análisis comparativo*. International Union for Conservation of Nature (UICN). https://portals.iucn.org/library/sites/library/files/documents/2006-065.pdf

Schmidt, G. (2010). *Territorios Indígenas y Áreas Protegidas*. Tres Casos y sus Conclusiones. Servicio de las Iglesias Evangélicas en Alemania para el Desarrollo (EED). https://ceccam.org/sites/default/files/Areas%20Protegidas.pdf

Secretaría de Medio Ambiente y Recursos Naturales (SEMARNAT). (2021). *Áreas Naturales Protegidas, piedra angular de la conservación*. Gobierno de México. Available on https://www.gob.mx/semarnat/articulos/areas-naturales-protegidas-piedra-angular-de-la-conservacion

Secretaría de Medio Ambiente y Recursos Naturales (SEMARNAT). (2011). *Biodiversidad. Conocer para Conservar. Serie ¿Y el medio ambiente?* Gobierno de México-SEMARNAT.

Secretariat of the Convention on Biological Diversity. (2008). *Protected Areas in Today's World: Their Values and Benefits for the Welfare of the Planet*. Secretariat of the Convention on Biological Diversity. Available on https://www.cbd.int/doc/publications/cbd-ts-36-en.pdf

The World Bank. (2010). *Valuing protected areas*. The World Bank. https://doi.org/10.13140/RG.2.1.2296.6642

Valverde Garnica, A. (2015). Gobernanza ambiental en áreas protegidas. El caso del Parque Nacional y Área Natural de Manejo Integrado Cotapata. *Revista Integra Educativa, 8*(3), 71–85. http://www.scielo.org.bo/pdf/rieiii/v8n3/v8n3_06.pdf

Vásquez Sánchez, M. (2017). Conservación de la naturaleza y áreas naturales protegidas en territorios de los pueblos originarios de la frontera sur de México. *Sociedad Y Ambiente, 15*, 117–130. https://doi.org/10.31840/sya.v0i15.1790

Chapter 4
Legislation and Management of Protected Natural Areas

Introduction

In addition to decrees establishing protected areas, other instruments are needed to contribute to the protection, conservation, and monitoring of these areas. Protected area decrees alone are not sufficient to achieve environmental conservation. Although increasing the number of protected areas is a valuable government initiative, it is not enough to have an "appropriate" number of areas in the right locations; it is also necessary to manage them effectively to achieve the desired results (Dearden et al., 2005, p. 98). Regardless of category or purpose, protected areas can only be effectively designated and safeguarded if there are interconnected political, administrative, and legislative instruments that contribute to the conservation and protection of these areas (Mackinnon et al., 1986; IUCN, 2005; Zielińska, 2010). Furthermore, the existence and functioning of human and financial resources are fundamental to the effective operation of protected areas (Pliego-Moreno et al., 2016).

Protected areas are confronted with challenges both within and on their periphery, with a high percentage of areas undergoing land use change, as well as various problems related to degradation and loss of biodiversity. These problems are associated with population growth, land use and natural resources, urbanisation and climate change. In order to address the aforementioned challenges, it is imperative to enhance the efficacy of protected area management (Bezaury-Creel & Gutiérrez Carbonell, 2009; IUCN, 2005; Ortega-Rubio et al., 2015). Historically, the classic model of governance for protected areas has been governmental; the land is property of the state, and decision-making is attributed to government agencies with the necessary responsibilities and operational competencies to create, regulate, and manage the different protected areas and applicable legislation (Lausche, 2012).

The objective of this chapter is to review and analyse the type of governance of protected areas and to examine the legislation and institutions related to federal

J. L. Carpio-Domínguez et al., *Drug Trafficking and Protected Areas in Mexico*,
https://doi.org/10.1007/978-3-032-18218-0_4

protected areas in Mexico. In addition, this chapter analyses the institutional capacity of the State to monitor and inspect protected areas through law enforcement within and on the periphery of protected areas. This will reveal any gaps or strengths in terms of the personnel authorised to enforce the law in protected areas.

Government Governance in Protected Areas

The law, policy, and administration of protected areas are important interconnected aspects that affect the successful management of protected areas. They merge to achieve objectives, effective planning, competent implementation, and the participation of society (Mackinnon et al., 1986). To achieve the objectives of protected areas, policy and administration must consider the interrelationships between society and nature, not only the ecosystemic dimension of the area but also its cultural dimensions and the interests and needs of the population that depends directly or indirectly on the resources within the areas (Balloffet & Martin, 2007, p. 2; Gravez et al., 2011).

There are instruments, institutions, and processes that can be influenced by those involved in protected areas, such as exercising authority or fulfilling responsibilities. Instruments may come from international law, conventions, standards, and good practices for conservation, especially international conventions that have been ratified at the national level, and may come from national policy and legislation, from policies, strategies, agreements, and plans ranging from the national constitution to sector-specific legislation; any of the instruments must consider the time and work required for restoration and monitoring activities, for enforcing decisions and regulations, accompanied by an active judiciary willing to impose fines and other sanctions (Borrini-Feyerabend et al., 2014).

Different categories of protected areas require different levels of protection and use (Mackinnon et al., 1986). For all categories of protected area management, the effectiveness of management is influenced by the quality of governance (Valverde Garnica, 2015). Governance in protected areas refers to the processes by which decisions are made and implemented (or not), including all official and unofficial actors involved in decision-making and implementation; it encompasses the official and unofficial structures established to make and implement decisions and includes policies, laws, decrees, regulations, instruments, institutions, processes, cultural values, and the ways in which responsibility and power are exercised; in other words, governance encompasses all the means, practices, and interactions between the who, why, and how involved in the establishment and management of protected areas in order to achieve the objectives of those areas (IUCN, 2005; Rivas Toledo, 2006; Gravez et al., 2011; Lausche, 2012).

It is possible to distinguish four types of governance in protected areas, depending on the actors who have the authority and responsibility for major decisions, such as implementing decrees, determining objectives, plans, and inspection and surveillance authority: (1) governance by the government (at different levels and possibly combining different institutions), (2) shared governance (e.g. joint governance between different rights holders and stakeholders), (3) governance by individuals and private organisations, and (4) governance by indigenous peoples and local communities (IUCN, 2005; Borrini-Feyerabend, 2007; Dudley, 2008; Day et al., 2012; Borrini-Feyerabend et al., 2014). The classic model of governance for protected areas that prevails throughout the world is governmental, which usually means that the land where protected areas are located is owned or controlled by the state, and it is the responsibility of the government and its agencies to make decisions to create and manage the areas in accordance with applicable legislation, which not only contains the objectives pursued by the law, but also serves as a framework for assessing whether certain actions and decisions taken by society are in compliance with the law (Lausche, 2012).

In government-managed protected areas, which vary by country, one or more government agencies typically have the authority, responsibility, and accountability to designate, manage, and monitor protected areas, determine their conservation objectives, develop and enforce their management plans, and often the state or federal government may also own the land, water, and related resources. The government retains overall control of a protected area and makes all important decisions, but delegates planning and/or day-to-day management tasks supervised by the government to other actors, such as subnational and municipal government agencies, private actors, or the community (Borrini-Feyerabend, 2007; Dudley, 2008; Day et al., 2012; Borrini-Feyerabend et al., 2014).

In Mexico's federal protected areas, governance by the government predominates, as it is the federal government, through its institutions, that decrees, authorises, monitors, and enforces the law. There are various types of protected areas in the country (state or municipal governments, or areas voluntarily designated for conservation by local communities), but it is the federal government's protected areas that are most recognised in national and international policy. The Ley General del Equilibrio Ecológico y la Protección al Ambiente (LGEEPA) (General Law of Ecological Balance and Environmental Protection), Mexico's main national environmental law, establishes that federal protected areas (biosphere reserves, national parks, natural monuments, natural resource protection areas, flora and fauna protection areas, sanctuaries, and areas voluntarily designated for conservation; the latter not being internationally recognised) fall under the jurisdiction of the Federation (Article 46 of the LGEEPA, 2024). Even the legitimate owners, possessors, or holders of other rights over land, water, and forests within protected areas must comply with the modalities established by law (Article 44 of the LGEEPA, 2024).

Protected areas must have plans, programmes, regulations, and management objectives with defined strategies to achieve the desired level of protection for each area (Costa Pellizzaro et al., 2015). Each protected area will have its own management needs, including the necessary monitoring, research, and support services, but one of the most important provisions that protected area legislation should establish as a requirement is a management plan or programme for each protected area designated by law, preferably once the area has been decreed (Mackinnon et al., 1986). The management, conservation, governing, etc., plans or programmes established for each area decreed in countries worldwide elaborate on and specify the conservation and use strategies for protected areas, how they should be managed, and establish specific restrictions on their use (Lausche, 2012).

These plans or programmes must be established once the area has been designated, but it is advisable to continuously carry out periodic reviews and evaluations. Some countries specify in their legislation the timeframes within which the plan or programme must be established and implemented, as well as the procedures for the adoption, approval, and review of the plan or programme (Mackinnon et al., 1986). In Mexico, protected area management programmes are instruments that determine conservation and use strategies for protected areas at the national and global levels.

Legally, programmes are defined as the guiding, planning, and regulatory instrument that establishes the activities, actions, and basic guidelines for the management and administration of the respective protected natural area (Article 3, Section XI, of the LGEEPA Reglamento de la LGEEPA en materia de ANP [Regulations on protected areas], 2014). The LGEEPA, Mexico's highest environmental law, states that the federal government shall formulate, within 1 year of the publication of the decree for the respective area, the management programme for the protected area (Article 65 of the LGEEPA, 2024).

Environmental Legislation on Protected Natural Areas in Mexico

Environmental public policy instruments include regulations such as legislation (Mickwitz, 2003). The existence of an adequate legal framework is fundamental to the success of any protected area. The regulation of activities related to protected areas can be of three types: (1) general, applicable within the protected area; (2) internal, depending on the protected area; and (3) applicable outside the protected area associated with external activities that could be harmful to the area. Legislation provides a general framework on how the area or different categories of protected areas should be protected. It should contain provisions on the preservation of biodiversity and nature and identify the purpose or objectives pursued by the law, which will serve for its subsequent application and evaluation of actions and decisions in accordance with the law, in the administrative and/or criminal sphere (Mackinnon et al., 1986; Lausche, 2012).

Legislation on protected areas may cover objectives related to compliance with obligations under international treaties or other legally binding instruments for the country (Lausche, 2012). It is advisable for national legal frameworks to include elements of the international agreements to which they are party, but they must also be in harmony with local traditions, institutions, and ecological conditions (Mackinnon et al., 1986). Even in countries with a constitutional framework, they can adopt international commitments or include their own so that there are constitutional provisions that assign issues and jurisdiction in issues related to protected areas (Mackinnon et al., 1986).

Each country can choose whether to incorporate international guidelines into its legislation governing protected areas, which has led to a variety of positions being adopted by many countries (Costa Pellizzaro et al., 2015). Most countries have national legislation relating to protected areas, but this varies from country to country, mainly because each has its own category system and the legislation incorporates these systems (Dudley, 2008). However, in general, legal provisions in protected areas stipulate controls (which will vary not only from country to country but also from area to area), establish the agency authorised to protect the area by law, and specify what can and cannot be done inside and outside the protected area. They help to control the activities conducted by people living in protected areas and visitors by inspecting permits for, for example, the sale of food and products, hunting, fishing, the use of wildlife and forest resources, the use of campfires, soil extraction, the use of vehicles, logging, agricultural activities, among others, and in the event of irregularities, applying sanctions (Mackinnon et al., 1986).

On 13 June 1992, Mexico signed the Convention on Biological Diversity (CBD), which entered into force in the country on 29 December 1993, and as part of its commitments in 2004, the country adopted the CBD's Programme of Work on Protected Areas (POWPA), whose main objective is to support the creation and maintenance of national and regional protected area systems, ensuring their effective management and representativeness (Secretaría del Convenio sobre la Diversidad Biológica, 2004; CONANP, 2016; Secretaría de Relaciones Exteriores, 2025). Bishop et al. (2004, p. 64) reviewed in 2004 the extent to which the IUCN category guidelines established in 1994 had been incorporated into countries' national legislation, finding that of a total of 126 national laws reviewed from 87 countries since 1994, 10.0% had used the international categories, with Mexico being one of those countries where it was observed that the proposals had a significant influence (Costa Pellizzaro et al., 2015).

In Mexico, since the early 1990s, a national environmental policy has gradually been developed with protection and conservation criteria that have helped to consolidate instruments such as protected areas. As part of its environmental policy, the country has cross-cutting administrative and criminal law that addresses measures to be complied with by society inside and outside protected areas. A national law on protected areas that incorporated the guidelines and commitments of international law into its legislation. In January 1988, the Ley General del Equilibrio Ecológico y la Protección al Ambiente (LGEEPA) (2024) was implemented in the country, considered the most important administrative environmental law in force, and since then the issue of protected areas has been included; and in February 2002, the Código Penal Federal (2024) included crimes related to this issue in criminal law. Table 4.1 presents the main national regulations that address the issue of protected areas in criminal and administrative matters at the federal level, either directly or transversally, as there are provisions on protected areas in regulations in other environmental areas.

Table 4.1 Regulations applicable to (federal) protected areas in Mexico until 2024

Regulations
Criminal
Código penal federal (Federal Criminal Code)
Administrative
Ley General del Equilibrio Ecológico y la Protección al Ambiente (LGEEPA) (General law on ecological balance and environmental protection)
Ley de Aguas Nacionales (LAN) (National Water law)
Ley General de Vida Silvestre (LGVS) (General wildlife law)
Ley General de Desarrollo Forestal Sustentable (LGDFS) (General law on sustainable Forest development)
Ley de Bioseguridad de Organismos Genéticamente Modificados (genetically modified organisms biosafety law)
Ley General para la Prevención y Gestión integral de los Residuos (LGPGIR) (General law for the prevention and Comprehensive Management of Waste)
Reglamento de la LGEEPA en materia de Áreas Naturales Protegidas (LGEEPA regulations on protected areas)
Reglamento de la ley de Aguas Nacionales (LAN) (regulations of the National Water law)
Reglamento de la ley General de Vida Silvestre (LGVS) (regulations of the General wildlife law)
Reglamento ley General de Desarrollo Forestal Sustentable (LGDFS) (regulations General law on sustainable Forest development)
Reglamento de la LGEEPA en materia de Evaluación del Impacto Ambiental (EIA) (LGEEPA regulations on environmental impact assessment)
Reglamento de la ley de Bioseguridad de Organismos Genéticamente Modificados (regulations of the law on biosafety of genetically modified organisms)
Reglamento de la ley General para la Prevención y Gestión integral de los Residuos (LGPGIR) (regulations of the General law for the prevention and Comprehensive Management of Waste)
Normas Oficiales Mexicanas (NOM) (official Mexican standards)
NOM-001-SEMARNAT-1996. Establece los límites máximos permisibles de contaminantes en las descargas de aguas residuales en aguas y bienes nacionales. (NOM-001-SEMARNAT-1996. Establishes the maximum permissible limits of pollutants in wastewater discharges into national waters and property)
NOM-059-SEMARNAT-2010. Protección ambiental-Especies nativas de México de flora y fauna silvestres-Categorías de riesgo y especificaciones para su inclusión, exclusión o cambio--Lista de especies en riesgo. (NOM-059-SEMARNAT-2010. Environmental protection—Native species of wild flora and fauna in Mexico—Risk categories and specifications for inclusion, exclusion, or change—List of species at risk)
NOM-152-SEMARNAT-2006. Establece los lineamientos, criterios y especificaciones de los contenidos de los programas de manejo forestal para el aprovechamiento de recursos forestales maderables en bosques, selvas y vegetación de zonas áridas. (NOM-152-SEMARNAT-2006. Establishes the guidelines, criteria, and specifications for the content of forest management programmes for the exploitation of timber resources in forests, jungles, and vegetation in arid areas)

Source: the authors

Criminal Jurisdiction

The Código Penal Federal (2025) (Federal Criminal Code) contains federal crimes, general rules on crimes and criminal liability, penalties and security measures, and rules on the application of criminal sanctions. Articles 414 to 423 of the Federal Criminal Code stipulate crimes against the environment and nature, including crimes against protected areas. The Federal Criminal Code penalises activities that cause harm to natural resources, flora, fauna, ecosystems, water quality, soil, subsoil, or the environment in protected areas, related to:

(a) Produce, store, transport, abandon, dispose of, discharge, or perform any other activity involving hazardous substances[1] (Article 414).
(b) Emit gases, fumes, dust, and pollutants into the atmosphere and generate noise, vibrations, thermal or light energy from federal sources (Article 415).
(c) Discharging wastewater, chemical or biochemical liquids, waste or pollutants into soil, subsoil, marine waters, rivers, basins, reservoirs, or other water bodies or watercourses under federal jurisdiction (Article 416).
(d) Removing or destroying forest vegetation, cutting, uprooting, felling or logging one or more trees, and changing land use without authorisation (section 418).
(e) Transport, trade, dispose of, distribute, supply, stockpile, purchase, receive, acquire, store, safeguard, possess, or transform forest raw materials or timber forest products without authorisation (Article 419).
(f) Capture, harm, or kill any specimen of turtle or marine mammal, or collect or store any of their products or by-products; capture, transform, collect, transport, or harm specimens of aquatic species declared to be protected; capturing, processing, stockpiling, transporting, destroying, or trading abalone, shrimp, sea cucumbers, and lobsters, inside or outside closed seasons, without authorisation; hunting, fishing, or capturing any specimen of wildlife using unauthorised means, or endangering the biological viability of a wild population or species; trafficking, capturing, possessing, transporting, collecting, introducing, extracting, or damaging specimens, their products or by-products and other genetic resources of a species of wild flora or fauna that is closed season, endemic, threatened, endangered, subject to special protection, or regulated by an international treaty to which Mexico is a party (Article 420) and.
(g) Damaging, draining, or filling wetlands, mangroves, lagoons, estuaries, or marshes; damaging reefs; introducing or releasing any specimen of exotic flora or fauna into the natural environment; causing a fire in a forest, jungle, natural vegetation, or woodland (Article 420 Bis).

[1] Hazardous substances are those with corrosive, reactive, explosive, toxic, flammable, radioactive, or other similar characteristics.

Administrative Jurisdiction

The main federal environmental law in Mexico was published in 1988 and remains in effective in 2025, the Ley General del Equilibrio Ecológico y la Protección al Ambiente (LGEEPA) (2024). It is a fundamental piece of legislation regulating relations between the government and the governed with the aim of preserving and restoring ecological equilibrium, including establishing the basis for the preservation and protection of biodiversity, as well as the establishment and administration of protected natural areas (Article 1, Section IV). It establishes that the Federation has the competence to establish, regulate, administer, and monitor protected natural areas under federal jurisdiction (Article 5, Section VIII). In addition, this Law establishes the definition of protected areas (Article 3, Section II) and the six categories, as well as the special regimes for protection, conservation, restoration, and development, which restrict access to and use of flora, fauna, and other resources in these areas, as well as water, soil, and air pollution within and around them, in order to ensure the balance and continuity of evolutionary and ecological processes. Information found in Articles 44 to 78 BIS 1.

The Ley de Aguas Nacionales (LAN) (2023) (National Water Law) regulates all issues relating to national waters (surface or underground) (rivers, lakes, springs, estuaries, and groundwater, among others), with the aim of regulating the exploitation, use, or utilisation of such waters, their distribution and control, as well as the preservation of their quantity and quality in order to achieve their comprehensive sustainable development (Articles 1 and 2). In protected areas, this law regulates the activities of extraction, discharge, and lighting, mainly of national waters within the areas. It also regulates the discharge of wastewater into natural soils within the areas. The activities covered by this law are regulated by the Regulations of the National Water Law (LAN) (2014).

The Ley General de Vida Silvetre (LGVS) (2021) (General Wildlife Law) aims to establish the concurrence and competences between the levels of government (federal, state, and municipal), as well as to regulate and administer issues related to the conservation and sustainable use of wildlife and its habitat in Mexican territory (Article 1). In protected areas, this Law regulates activities related to the use, extraction, introduction, hunting, subsistence, abuse, and trafficking of wildlife (flora and fauna) within the protected areas. The activities covered by this Law are regulated by the Regulations of the General Wildlife Law (LGVS) (2014).

The Ley General de Desarrollo Forestal Sustentable (LGDFS) (2024) (General Law on Sustainable Forest Development) regulates and promotes the conservation, protection, restoration, production, management, cultivation, and use of the country's forest ecosystems and resources, as well as distributing forestry-related powers as appropriate (Article 1). In protected areas, it regulates all activities involving the use of timber and non-timber forest resources, including reforestation and forest closures, when these are intended to be carried out wholly or partly in areas under the jurisdiction of the Federation (Article 61 Bis). The activities to be regulated include changes in the use of forest land to other types of land (agricultural, urban, and industrial). The activities covered by the Law are regulated by the Regulations of the General Law on Sustainable Forest Development (LGDFS) (2014).

The Ley de Bioseguridad de Organismos Genéticamente Modificados (2022) (Law on Biosafety of Genetically Modified Organisms) regulates, among other things, the experimental release of Genetically Modified Organisms (GMOs) in pilot or commercial programmes, with the aim of preventing, avoiding, or reducing the risks that these activities may pose to the environment and biological diversity (Article 1). In protected areas, activities involving GMOs in protected natural areas constitute administrative offences (Article 119, section XIII). Furthermore, activities involving GMOs are only permitted for bioremediation, and activities involving GMOs are prohibited in the core zones of protected natural areas (Article 89). Activities to be regulated include changes in land use from forest to other types of land (agricultural, urban, and industrial). The activities covered by the Act are regulated by the Regulations of the Act on Biosafety of Genetically Modified Organisms (2009).

The Ley General para la Prevención y Gestión Integral de los Residuos (2023) (General Law for the Prevention and Comprehensive Management of Waste) regulates the generation, recovery, and management of hazardous, mining and metallurgical waste, municipal solid waste, and waste requiring special handling to prevent site contamination (Article 1). In protected areas, this Law prohibits the discharge and final disposal of hazardous waste. In addition, this Law prohibits the final disposal of mining and metallurgical waste in protected areas (Article 17). The activities covered by this Law are regulated by the Regulations of the General Law for the Prevention and Comprehensive Management of Waste (2014).

The LGEEPA Regulations on Environmental Impact Assessment (EIA) (2014) aim to regulate the LGEEPA in matters of environmental impact assessment[2] at the federal level. In protected areas, the regulations establish that any type of work or installation to be conducted within protected areas under the jurisdiction of the Federation shall require prior authorisation from the federal government, with the exception of: (a) Subsistence and domestic use activities, provided they are carried out by communities settled in the area; (b) activities that are essential for the conservation, maintenance, and surveillance of the areas; (c) urban infrastructure works and housing development in urbanised areas located within the areas and not prohibited by legal provisions, and (d) construction of dwellings on agricultural or livestock land or within the boundaries of existing population centres, when located in rural communities (Article 5, paragraph S).

The Regulations of the General Law of Ecological Balance and Environmental Protection in relation to Protected Natural Areas (2014) establish everything related to the establishment, administration, management, and sanctions of protected areas under the jurisdiction of the Federation (Article 1). It includes aspects that must be adopted for the administration of protected areas: guidelines, institutional

[2] An environmental impact assessment is a government procedure conducted to identify and mitigate the effects that a new project may have on the environment during its site preparation, construction, operation, and abandonment stages.

mechanisms, programmes, policies, and actions aimed at the conservation, preservation, protection, and restoration of ecosystems, the sustainable use and exploitation of natural resources, and inspection and surveillance; measures related to financing for their operation, instruments to promote coordination between levels of government and concerted action with the public, social and private sectors; and actions to promote the training and education of technical support personnel (Article 5).

In order to visualise what has been achieved in Mexico in terms of developing the necessary capacities for managing its protected areas, it is necessary to understand the legal framework that underpins them, but it is also important to understand the institutional framework for their administration, the human and financial resources allocated to their conservation and law enforcement, among other factors (Carabias et al., 2003; Bezaury-Creel & Gutiérrez Carbonell, 2009). For protected areas to be effective, it is necessary, among other things, to monitor compliance with management regulations, with inspection and surveillance being a core aspect that prevents, monitors, and acts on activities in these areas (de la Garza Buentello, 2014; Henson et al., 2016; Wyatt et al., 2020).

Normas Oficiales Mexicanas (NOM's) (Official Mexican Standards)

The administration of protected areas must be conducted in accordance with the provisions of laws, regulations, management programmes, and Normas Oficiales Mexicanas (NOMs). The Normas Oficiales Mexicanas (NOMs) are technical regulations created with scientific and technological information, issued by competent authorities and containing mandatory compliance parameters. In protected areas, the use of wild and aquatic flora and fauna and forest resources within the areas must be conducted in accordance with the restrictions contained in the NOMs. For example, activities for the conservation of ecosystems and their elements, scientific research and ecological education, and the use of wildlife and forest resources, including those for research and experimentation purposes, shall be subject to the restrictions established in the NOMs. In addition, the use, exploitation, and utilisation of national waters located in protected areas shall be subject to the NOMs, as well as to regulations designed to prevent water and soil contamination within the areas.

The NOMs on environmental issues include topics such as the atmosphere, water, soil, waste, wildlife, and forest resources. Although there is no specific standard on protected areas, the topics are cross-cutting, and some of them may apply to protected areas depending on the topic. Among the main NOMs applicable to protected areas in Mexico is NOM-001-SEMARNAT-2021, which contributes to establishing the maximum permissible limits of pollutants in wastewater discharged into national assets (rivers, lakes, lagoons, estuaries, marine areas and other national waterways and natural soil). NOM-059-SEMARNAT-2010 contributes to identifying species or populations of wild flora and fauna at risk (probably extinct in the wild, endangered, threatened, and subject to special protection) and endemic species within protected areas in our country for appropriate care and protection, through the integration of the corresponding lists. In another example, NOM-152-SEMARNAT-2006 contributes to establishing the guidelines, criteria, and specifications for the content of forest management programmes for the use of timber resources in forests, jungles, and vegetation in arid areas, including those that may apply within protected areas.

Institutions in the Governance of Government-Protected Areas in Mexico

Although there have been significant changes in relation to protected areas, the lack of instruments for administration and conservation remains one of their historical problems (Simonian, 1999). The existence of policy and legislation is fundamental to the success of any protected area programme, but agencies or authorities authorised to protect areas by law are also indispensable; even robust laws need to be enforced by competent actors authorised to support effective performance through law enforcement (Mackinnon et al., 1986; Kakira, 2010; Zielińska, 2010; Lausche, 2012). In a classic protected area, the State makes decisions through a government agency vested with the necessary operational responsibilities and competencies to legislate, create, and manage the different areas in accordance with the relevant legislation (Lausche, 2012). A prerequisite for effective legislation on protected areas is the proper allocation of authority and responsibilities for monitoring and enforcement, specifying who has the authority to act as the authority and stipulating the functions, responsibilities, and competencies of that authority (Mackinnon et al., 1986; Lausche, 2012; Henson et al., 2016; Wyatt et al., 2020).

The actors involved in the governance of protected areas are varied. On the government side, they may include legislators, agencies responsible for protected area systems, agencies and authorised officials responsible for administering and enforcing the law, government agencies and officials responsible for and interested in exploiting natural resources (e.g. water, agriculture, forestry, and fisheries), law enforcement authorities, tribal governments, and designated local authorities (Borrini-Feyerabend et al., 2014). In any protected area, it is important that there be an authority responsible for law enforcement, or that there be actors who exercise that function (such as rangers or surveillance committees) who have a working relationship with the authority responsible for law enforcement in order to control abuses against protected areas (Mackinnon et al., 1986).

The responsibilities of the actors depend on the needs of the protected area, the faculties, skills and strengths of the officer and the institutions; faculties may include police authority to detain, search, and arrest in accordance with the Penal Code,

Table 4.2 Main institutions in (federal) protected areas in Mexico

Institutions
Criminal jurisdiction
Fiscalía General de la República (FGR) (attorney General's office)
Administrative jurisdiction
Secretaría de Medio Ambiente y Recursos Naturales (SEMARNAT) (secretariat of environment and natural resources)
Comisión Nacional de Agua (CONAGUA) (National Water Commission)
Comisión Nacional de Áreas Naturales Protegidas (CONANP) (National Commission for protected areas)
Procuraduría federal de Protección al Ambiente (PROFEPA) (Federal Attorney General's Office for Environmental Protection)
Comisión Nacional Forestal (CONAFOR) (National Forestry Commission)
Comisión Nacional para el Conocimiento y Uso de la Biodiversidad (CONABIO) (National Commission for the knowledge and use of biodiversity)
Public security
National Guard
State and municipal police forces
State and municipal agencies
Environmental agencies and state attorney General's offices

Source: the authors

other faculties are administrative, including surveillance and inspections to apply sanctions, or may include educative, training, and guidance functions; in all cases, these actors are often the ones who have the most contact with local communities (Lausche, 2012). In Mexico, there are various actors and institutions involved in government governance in protected areas, ranging from criminal, administrative, and public security actors, with objectives related to the conservation and protection of natural resources in protected areas and to surveillance and law enforcement. The institutions, their faculties and obligations related to federal protected areas in Mexico are presented in Table 4.2. Each institution has specific functions that they implement at different spatial levels within and on the periphery of protected areas.

Institutions in Criminal Jurisdiction

The Fiscalía General de la República (FGR) (Attorney General's Office) is an autonomous constitutional agency whose purpose is to investigate crimes punishable under criminal law and clarify the crimes by providing effective and efficient justice in accordance with the law, thereby helping to combat and reduce insecurity and prevent crime (FGR, 2024; Ley de la Fiscalía General de la República, 2024). Its objectives include investigating and punishing acts that the law defines as crimes against the environment under the Federal Criminal Code which, as noted in the previous section, include those related to protected areas. To this end, it has a unit known as the Special Prosecutor's Office for the Investigation of Crimes against the Environment and Crimes under Special Laws (FGR, 2024).

Institutions in Administrative Jurisdiction

The Secretaria de Medio Ambiente y Recursos Naturales (SEMARNAT) (Secretariat of Environment and Natural Resources), created in 2000, is considered the main federal institution with the authority to administer protected areas. It is the authority responsible for decreeing and designating protected areas, authorising management programmes, and granting authorisations for activities that may have an impact within and on the periphery of the areas. Although protected areas may be managed by interested individuals or legal entities, they must have technical, financial, or management capacity (Articles 6 and 7 of the Reglamento de la LGEEPA en materia de Áreas Naturales Protegidas, 2014). This makes it difficult in practice, which is why the 232 federally designated protected areas are managed by SEMARNAT. The Secretariat is supported by decentralised federal administrative institutions such as the National Commission for Protected Natural Areas (CONANP), the National Forestry Commission (CONAFOR), and the Federal Attorney General's Office for Environmental Protection (PROFEPA).

SEMARNAT delegates to CONANP, founded in 2000, attributions related to: (a) promoting the establishment, modification, recategorisation, or extinction of federal protected areas; (b) administering protected areas; (c) promoting and developing activities aimed at the conservation of ecosystems and their biodiversity in protected areas; (d) formulating, promoting, executing, and evaluating projects for the conservation and recovery of species and populations considered to be priorities; (e) involving society in the protection, preservation, restoration, and administration of the areas; (f) supporting surveillance tasks in the areas; among others (Article 81 of the Reglamento Interior de la Secretaría de Medio Ambiente y Recursos Naturales (SEMARNAT), 2025).

CONAFOR, created in 2001, has among its objectives to develop, promote, and encourage the use, protection, conservation, and restoration of all forest ecosystems, including those applicable to protected areas. It is also responsible for participating in the formulation of plans and programmes and in the implementation of sustainable forest development policy. It is also an internationally recognised organisation for its work in combating wildfires.

Since 1992, PROFEPA has been responsible for enforcing compliance with environmental legislation within and outside federal protected areas (Article 137 of the Reglamento en materia de Áreas Naturales Protegidas; Articles 50 and 52 of the Reglamento Interior de la SEMARNAT, 2025). As part of its inspection attributions, PROFEPA verifies compliance with regulations and takes action in protected areas to combat the illegal exploitation of natural resources, both extractive and non-extractive, by verifying compliance with the conditions contained in the corresponding authorisations. PROFEPA is therefore responsible for monitoring and inspecting the country's 232 protected natural areas, covering a total of 980007.19 km^2 of protected areas (terrestrial and marine) until 2024.

The Comisión Nacional del Agua (CONAGUA) (National Water Commission), founded in 1989, is responsible for verifying, inspecting, and checking compliance with the obligations set out in the Law, its Regulations, Official Mexican Standards, and other applicable provisions in relation to the exploitation, use, or utilisation of national waters, their inherent public assets, the extraction of stone materials, the discharge of waste water into national water bodies, and other national assets administered by this Commission (Articles 13, 14, 24, and 64 of the Reglamento Interior de la CONAGUA, 2012). Protected areas that safeguard sources of water are supervised by CONAGUA to ensure the conservation and protection of water.

The Comisión Nacional para el Conocimiento y Uso de la Biodiversidad (CONABIO) (National Commission for the Knowledge and Use of Biodiversity), created in 1992, has the objective of promoting, coordinating, supporting, and implementing activities aimed at understanding biological diversity, as well as its conservation and sustainable use, to provide data, information, and advice, as well as to implement national and global information networks on biodiversity (CONABIO, 2024). In protected areas, this Commission contributes to conservation, education, and research activities conducted therein, generating documentary, virtual, and cartographic information on federal protected areas.

Although there are other federal environmental institutions, they have limited or no direct relationship with protected areas, such as the Mexican Institute of Water Technology (IMTA), which acts as the technological arm of the water sector, with responsibilities related to human resources training to strengthen institutional capacities; the National Institute of Ecology and Climate Change (INECC), which is a research organisation focused on generating and integrating technical and scientific research on ecology and climate change issues for decision-making; and the National Agency for Industrial Safety and Environmental Protection in the Hydrocarbons Sector (ASEA), which regulates and supervises industrial safety, operational safety, and environmental protection with regard to activities in the hydrocarbons sector.

Public Security Institutions in Protected Areas

The National Guard is a civil public security institution, attached as a decentralised administrative body to the Secretaría de Seguridad y Protección Ciudadana (Secretariat of Security and Citizen Protection) (Article 4 of the Ley de la Guardia Nacional [National Guard Law], 2023). The objectives of the National Guard include performing the public security function under the responsibility of the Federation and, in some cases, collaborating in public security tasks that correspond to the federal entities or municipalities (Article 5 of the Ley de la Guardia Nacional,

2023). In protected areas, the National Guard's objectives include safeguarding the life, integrity, security, property, and rights of individuals, as well as preserving liberties, safeguarding the Nation's property and resources, and implementing collaborative and coordinated actions with federal entities and municipalities (Article 6, Sections II, III, and IV, of the Ley de la Guardia Nacional, 2023). In addition, its attributions and obligations include collaborating with other federal authorities in surveillance, verification, and inspection functions conferred upon them by other laws (Article 9, Section XXV, of the Ley de la Guardia Nacional, 2023). In this regard, the National Guard supports surveillance and inspection tasks in protected areas. In 2016, the Environmental Gendarmerie was established as a unit of the former Federal Police (now the National Guard) of Mexico, a security force trained in the protection of ecosystems and species at risk, more specifically responsible for protecting the country's protected areas, but it was discontinued in 2020.

In addition to the National Guard, other state or municipal public security agencies are responsible for supporting the competent federal environmental authority in the tasks required. For example, in order to conduct administrative actions and procedures related to investigation, inspection, and surveillance, PROFEPA may request the assistance and support of Federal Public Administration agencies dedicated to homeland security, as well as the authorities of the states, municipalities, and territorial divisions of Mexico City, in accordance with applicable legal provisions or coordination agreements (Article 55 of the Reglamento Interior de la SEMARNAT, 2025).

However, PROFEPA may not only request support from any of the public security authorities of the three levels of government (federal, state, and municipal), it can also request assistance in the exercise of its attributions from the personnel of the agencies or entities of the Federal Public Administration, the federal entities, municipalities, and territorial divisions of Mexico City who, in terms of the applicable legal provisions and agreements, are accredited as federal inspectors or investigators (Article 55 of the Reglamento Interior de la SEMARNAT, 2025).

Strong and competent institutions to manage, enforce laws and monitor compliance with protected area plans or programmes are crucial. It is important that the protected area administration has actors responsible for enforcing the law. Authorised officers should have the status of public servants when acting in the performance of their duties in accordance with the law (Lausche, 2012).

Law Enforcement in Protected Areas

The effectiveness of public administration cannot be measured based on the number of regulations approved, but rather on the enforcement of laws (de Montalvo Jääskeläinen, 2016, p.7). For legislation to "work", it must not only be well designed, but also effectively implemented and enforced; poor enforcement can affect policy effectiveness (Tremper et al., 2010; Gunningham, 2011). Law enforcement is a reactive policy instrument used by governments to investigate whether proactive policy instruments are achieving the expected results and to promote the achievement of objectives (Johansson, 2010; Monk, 2012). Law enforcement authorities monitor and inspect the environmental impacts, operations, permits, conditions, and obligations of social actors to assess whether they comply with the law; enforcement must be continuous (Nilsson, 2011, p. 619).

Inspection and surveillance actions are frontline actions and are important factors in deterring illegal activities in a protected area, and enforcement personnel are a key resource for maximising the impact of law enforcement in the field (Correa, 2016; Henson et al., 2016; Wyatt et al., 2020). Surveillance tends to focus more on maintaining a presence in protected areas, while inspection is the process of verifying compliance with the legal framework through law enforcement. Both can be implemented by authorised personnel from agencies with competence in this area (Pliego-Moreno et al., 2016).

In a broader sense, inspection verifies compliance with the legal framework in order to ensure compliance with laws, regulations, decrees, standards, permit and authorisation conditions, etc., in order to, among other things, protect natural resources and ecosystems in protected areas; while surveillance consists of maintaining a presence at sites of interest in order to prevent illegal activities (Lausche, 2012; Pliego-Moreno et al., 2016). In addition, inspections are conducted not only to identify specific offenders and bring them back into compliance, but also to deter offenders and all regulated entities in similar conditions of non-compliance through civil, administrative, or criminal proceedings (OECD, 2004; Johansson, 2010; Rose, 2011, p. 9).

The need for law enforcement is implicit behind the creation and use of law to protect protected areas (Mackinnon et al., 1986). Law enforcement is an important and necessary strategy in protected area issues worldwide, necessary to achieve conservation goals and objectives by monitoring compliance with permissions and conditions (Alers et al., 2007; Stab & Henle, 2009; Nilsson, 2011; Lausche, 2012; OECD, 2012; Borrini-Feyerabend et al., 2014; Pliego-Moreno et al., 2016). Monitoring and inspecting areas involve routine checks within the areas, verifying boundaries and, sometimes, monitoring outside the protected area, as well as visiting both the site and local people to monitor changes and collect data on the availability of natural resource and wildlife management. In addition, inspection includes checking that the appropriate permits are in place and that area regulations are not being violated, that only authorised personnel are active in restricted areas, and that there is no illegal hunting, logging, or deforestation for agriculture (Mackinnon et al., 1986; Henson et al., 2016; Wyatt, 2015). Therefore, inspection and surveillance efforts are essential to prevent, monitor, and respond to irregular human activities. Surveillance and inspections of protected areas must be adjusted to the existing natural, social, and economic realities and, at the same time, must be a conservation strategy for these areas (Zielińska, 2010).

Government institutions and processes are key to ensuring law enforcement and compliance, particularly in developing countries where there is a critical need to strengthen environmental legislation and institutions (Hongjun et al., 2017). Activities depend on a solid foundation of institutional competencies and functions

(Lausche, 2012; Henson et al., 2016). For inspection purposes, it is essential that this be performed by personnel authorised by the authorities with such attributions, while surveillance may be conducted by personnel authorised by agencies with expertise in the matter, by other institutions with similar faculties (water, forestry, wildlife, etc.), or even by members of community organisations trained and authorised by the responsible agencies (Pliego-Moreno et al., 2016). In other words, while surveillance can be conducted by both authorities and citizens, inspection can only be conducted by the authorised authority.

Thus, the responsibilities of authorised officers take a variety of forms, depending on the needs of the protected area and the skills and strengths of the officer, with common enforcement responsibilities including detention, search, and arrest in accordance with the Criminal Code and applicable rules of criminal procedure (Lausche, 2012). Authorised officials are generally civil servants who work on the front line because they deal directly with people in their roles as citizens, customers, and representatives of different organisations and operational activities; a group of public employees who position themselves "between" civil society and the State (Johansson, 2010; Monk, 2012). They are the individuals on both sides of regulation, the regulator and the regulated, who work daily to turn environmental regulation into action (Pautz, 2009). They are therefore in an intermediate position where their work is judged both from above (political and administrative management) and from below (operational activities and citizens) (Johansson, 2010; Lausche, 2012).

People who conduct environmental inspections are also often referred to as "street-level bureaucrats" as they are government employees who form an essential part of the political bureaucracy, but they are more often end users of policies than collaborators in their development, and they spend most of their time in the field, working directly on law enforcement, and play catch-up on policy reviews they regularly receive from the agency level (Pautz & Schnitzer, 2008; Monk, 2012).

In practice, the optimal number of officers for effective surveillance and law enforcement depends on a variety of factors, including the size and location of the area, the type of threat faced, human population density, and the animal species requiring protection. Some law enforcement professionals recommend that patrols and inspections should consist of at least two officers and at least one officer per 10–50 km^2, depending on needs (Mackinnon et al., 1986; Henson et al., 2016).

In another example, the SADC Regional Rhino Conservation Program recommends one park ranger per 20 km^2 in sanctuaries, while other studies based on rhino poaching recommend densities of up to one ranger per 9–19 km^2 and one forest ranger per 10 km^2; in another case, a density of one park ranger per 50 km^2 has been suggested for the protection of elephants; however, another study estimated that the optimal density of park rangers to protect elephants, i.e., when zero elephants are killed, is approximately one park ranger per 24 km^2 (Henson et al., 2016).

However, it is not simple to suggest a number of personnel to supervise protected areas; each area needs to identify its own number of personnel based on its specific challenges. For example, areas where the main threat is agricultural encroachment may not require as high a density of rangers as those experiencing intense poaching

(Henson et al., 2016). Furthermore, staffing levels are often beyond the control of line managers, as it depends on a variety of factors related to institutional management, personnel management, funding allocation, or national policies, among other issues (Lausche, 2012; Henson et al., 2016).

It has been demonstrated that as effective surveillance and law enforcement activities increase, there are fewer illegal activities within the protected area, and the number of inspectors is the factor most strongly correlated with effectiveness (Alers et al., 2007; Henson et al., 2016). Analysis of law enforcement contributes to better deployment of personnel, provides information on the efforts of those responsible, and makes visible the consequences, positive or negative, of political efforts (Mackinnon et al., 1986; Tremper et al., 2010; García-Escudero, 2015, p. 186).

Evaluating government agencies in protected areas illustrates the scope, capacities, constraints, and needs for strengthening and action by governments to achieve conservation objectives and the development of protected areas (Rivas Toledo, 2006; Valverde Garnica, 2015). The government's ability to enforce regulations in protected areas is often limited by social conflicts between stakeholders, lack of resources, budget, management programmes, and personnel for conservation and monitoring. Frequently, administrators and other frontline actors do not have sufficient knowledge, skills, capacities, and resources to ensure that protected areas respond to the challenges posed by global changes (IUCN, 2005; Brenner, 2010; Lausche, 2012; Correa, 2016). In protected areas, it is important to analyse not only the type of governance (who governs), but also its quality (how it is carried out) (Lausche, 2012).

Among the recommendations for more effective surveillance and inspection is to cover the entire protected area and not just check the boundaries, as potential offenders may be "invisible" unless surveillance is thorough, constant, and performed at irregular intervals, on different routes, every day, both night and day (Mackinnon et al., 1986).

Enforcement of Environmental Law in Mexico

Legislation on protected areas not only sets out the objectives pursued by the policy, but also serves as a framework for assessing whether certain actions and decisions taken by society are in accordance with the law (Lausche, 2012). Organisations, conservationists, and society in general celebrate the increase in the size of protected areas and the greater focus on conserving biodiversity, but they disagree on how protected areas should be managed (Naughton-Treves et al., 2005). In Mexico, problems have been identified in the country's protected areas, associated with vegetation and soil transformation, species loss and introduction, among others (Ortega-Rubio et al., 2015; CONANP, 2020), in addition to an insufficient number of inspectors to enforce the law (Castro-Salazar & Camacho, 2020).

For example, problems within protected areas in Mexico, in the Monarch Butterfly Biosphere Reserve, included increases in the rate of introduction of non-native species, erosion on trails and the opening of new trails, a lack of organisation, and accumulation of waste (Ortega-Rubio et al., 2015). In the Sierra Gorda Biosphere Reserve, areas of tropical forest and vegetation are shrinking; in the Sacromonte National Park, there has been a reduction in surface area due to the invasion of irregular human settlements; in the Ciénegas del Lerma Flora and Fauna Protection Area, species of flora and fauna have become extinct, and several endemic species are threatened or endangered (Arriola-Padilla et al., 2015). In the Upper Gulf of California and Colorado River Delta Reserve, the intervention of agencies such as PROFEPA has failed to reduce poaching, which represents a percentage similar to or greater than that of legal fishing, making evident that the intervention of the authorities has not had a significant impact on resolving this issue (Rodríguez-Quiroz et al., 2015).

PROFEPA is responsible for inspecting and monitoring compliance with legal provisions applicable to protected areas, such as authorisations for exploration, exploitation, or use of resources, or other activities (Article 137 of the Reglamento de la LGEEPA en materia de ANP, 2000). To facilitate the exercise of its attributions, PROFEPA has administrative units in all 32 federal entities, with at least one representative office per state (Article 41 of the Reglamento interior de la SEMARNAT, 2025). Each representative office has within its territorial jurisdiction the attributions to order and conduct visits or inspection operations to verify compliance with the legal provisions applicable to the restoration of natural resources, the preservation and protection of forest resources, wildlife and aquatic species at risk, their ecosystems and genetic resources, the biosafety of genetically modified organisms, exotic species that threaten ecosystems, habitats, or species, and the use and exploitation of protected areas, among others (Article 80 of the Reglamento Interior de la SEMARNAT, 2025).

There is a PROFEPA Environmental Protection Representative Office (ORPA) in each of the 32 states (Fig. 4.1). Most of them are located in the main urban centres of the states, but they are mostly far from protected areas (Castro-Salazar & Camacho, 2020), which limits the federal administrative authority's ability to respond efficiently to surveillance and law enforcement.

In each office, PROFEPA has federal inspection and administrative investigation personnel who have the authority to act in inspection and surveillance activities, and others designated in the mandates or commission orders issued by the Director of PROFEPA and the persons in the offices of the Deputy Attorney General, the Director Generals, and the ORPA director in accordance with the applicable legal provisions (Articles 52, Section I, and 80 of the Reglamento Interior de la SEMARNAT, 2025).

PROFEPA schedules, organises, and conducts visits or inspection operations to monitor and evaluate compliance with legal provisions applicable to the use and exploitation of protected areas in various areas (wildlife, soil and forest resources) and to supervise actions that cause pollution within and on the periphery of the area (such as soil and atmospheric pollution). PROFEPA inspectors are primarily responsible for verifying compliance with legislation related to protected areas in the field.

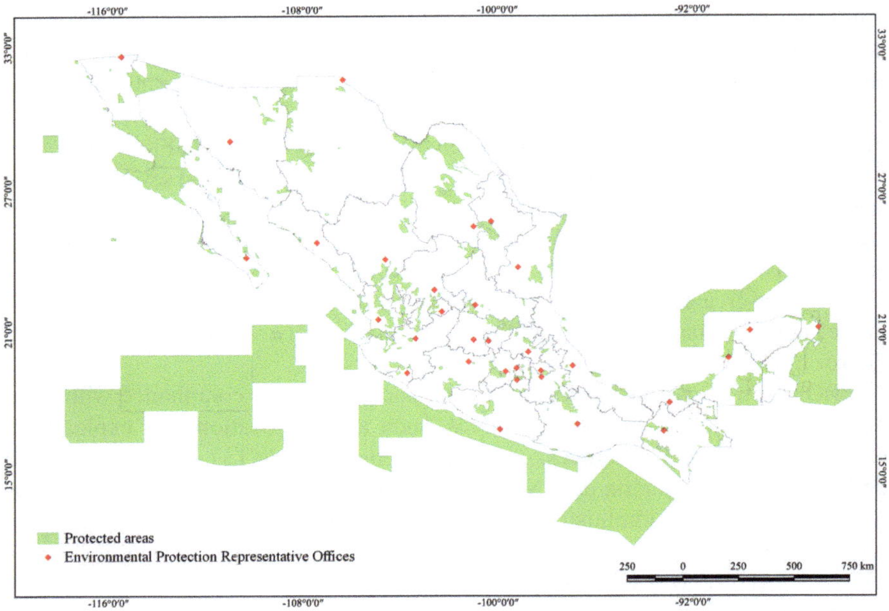

Fig. 4.1 Geographical location of PROFEPA Environmental Protection Representative Offices and protected areas in Mexico until 2024. **Source:** the authors

In 2024, there were a total of 470 inspectors nationwide to monitor and enforce the law in the 232 designated protected areas, an average of two inspectors per protected area. In 2015, approximately five inspectors were reported to be assigned per protected area, which is considered insufficient (Pliego-Moreno et al., 2016), and as a result, the number of inspectors per protected area decreased. Mexico City ($n = 61$) has the highest concentration of inspectors, as it is the state where PROFEPA's headquarters are located, followed by Chihuahua ($n = 24$); while Aguascalientes had the lowest number of inspectors in 2024 ($n = 1$) (Fig. 4.2). The Consejo Nacional de Evaluación de la Política de Desarrollo Social (CONEVAL) (2014) establishes that PROFEPA not only has the obligation to manage protected areas, but also to monitor and inspect forests and jungles, coastlines, and the conservation of wildlife and environmental pollution throughout the country, given its legal responsibilities.

In criminal justice issues, there was the Environmental Gendarmerie, through the now -extinguished Federal Police, created to prevent crimes and administrative offences in environmental issues, especially in the country's protected areas under a security and police operation model. When it was created in 2016, it had 192 gendarmes, and in its last year (2019), it had 232 environmental gendarmes. The deployment of a specialised unit for the prevention and response to environmental crimes in protected areas, from a public security perspective, aimed to provide tactical, response, and intelligence capabilities.

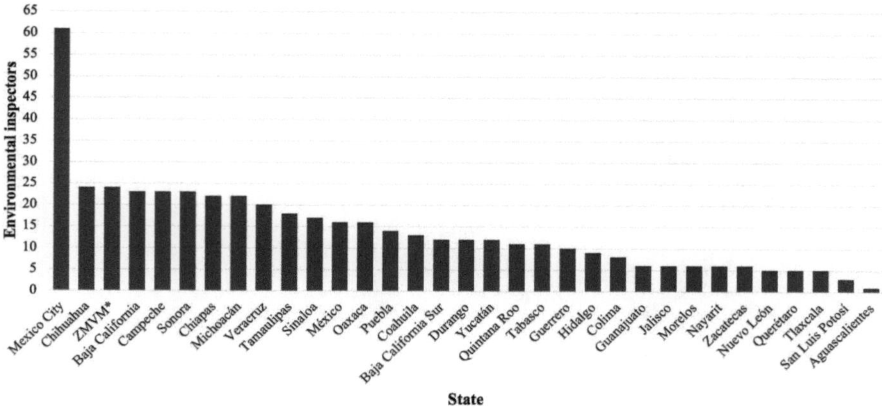

Fig. 4.2 PROFEPA environmental inspectors by state in Mexico until 2024. **Source:** the authors. *The ZMVM (Metropolitan Zone of the Valley of Mexico) includes some inspectors from Mexico City and some municipalities in the Estado de Mexico and one municipality in Hidalgo

When considering the land area of protected areas, Baja California (30,234.0 km^2) and Baja California Sur (25,912.2 km^2) are the two states with the largest land area of protected areas, but only Baja California stands out for having the highest average number of inspectors in 2024 ($n = 23$), while Baja California Sur has 12 inspectors (Fig. 4.3). Moreover, two of the four states with the smallest land area of protected areas, Sinaloa and Mexico City/ZMVM, 932.6 km^2 and 116.6 km^2, respectively, have more inspectors than states with larger protected areas, 17 and 85, respectively. Figure 4.3 shows that there is no correlation between the number of inspectors and the land area of protected areas per state; states with smaller areas have higher rates of inspectors.

When analysing the territorial extension of protected areas per inspector, by state, in Baja California Sur and Coahuila, each inspector must cover a larger area to monitor and enforce the law, 2159.4 km^2 and 1791.1 km^2 per inspector, respectively (Fig. 4.4). The large territorial extension to be monitored per inspector makes it difficult to cover a larger territorial extension in less time, which in turn has an impact on the effective enforcement of the law in protected areas.

Additionally, it is necessary to add the marine area to be inspected (see Fig. 4.1) in coastal states where marine protected areas are located, but this could not be presented in the analyses in Figs. 4.3 and 4.4 because the state or states responsible for monitoring and inspecting each protected area were not specified. The number of personnel, frequency, and intensity of surveillance and inspection between protected areas should vary, taking into account not only the size and location of the area, but also the type of threat, human population density, and the number and type of species requiring protection. For example, if it is not possible to increase the number of inspectors in all areas, at least increase surveillance and inspections in areas with protected species.

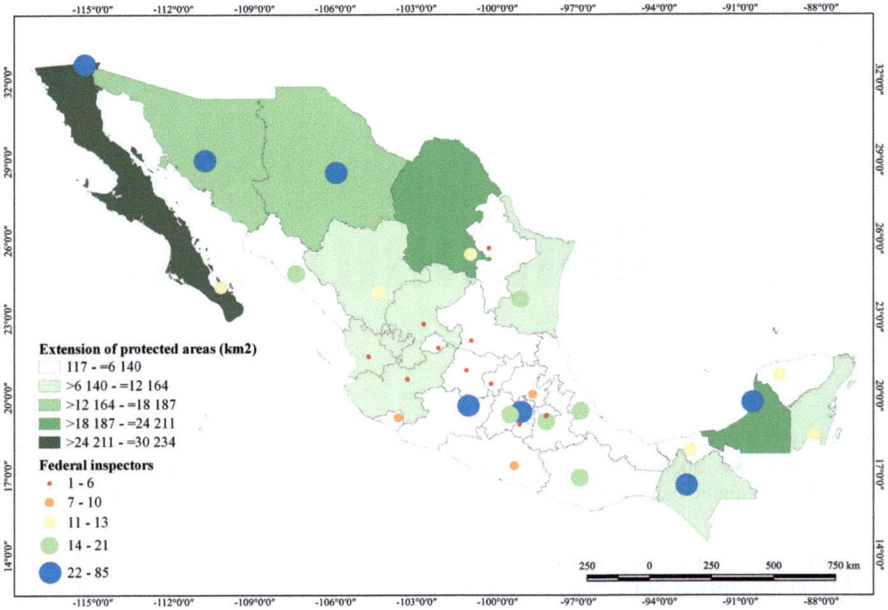

Fig. 4.3 Distribution of environmental inspectors and territorial extension of protected areas (km2) by state in Mexico until 2024. **Source:** the authors

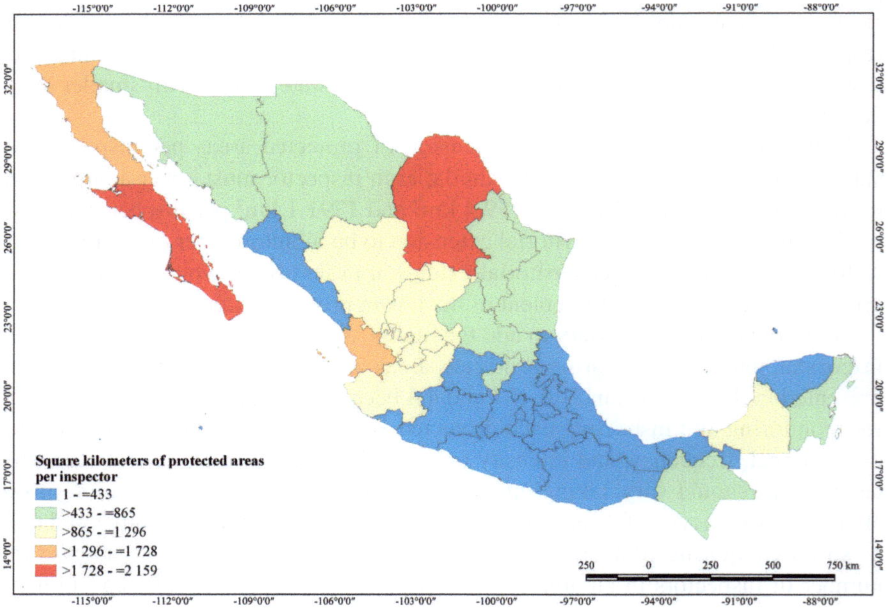

Fig. 4.4 Territorial extension of protected areas (km^2) per inspector by state in Mexico until 2024. **Source:** the authors

Another institutional limitation, in addition to the small number of offices and inspectors, is the budget allocated to environmental agencies. An analysis of the federal budget allocated to institutions directly involved in the administration and monitoring of protected areas shows that most of the federal budget goes to National Defence (which includes the army and air force) and Environment and Natural Resources (Fig. 4.5). Meanwhile, the Attorney General's Office, as the institution that prosecutes crimes in protected areas, receives the least amount of federal resources, along with the Public Security sector, despite being one of the first responders to crimes and administrative offences in protected areas.

When analysing and disaggregating the federal budget for the Environment and Natural Resources sector, the three main federal administrative institutions with the greatest responsibilities in protected areas are: SEMARNAT, PROFEPA, and CONANP (see sect. 3.2). The authority allocated the largest share of the total budget for this sector was SEMARNAT, with an average of 7.6% of the total between 2008 and 2024, followed by CONANP (2.2%), while the authority responsible for law enforcement was allocated 2.0% of the average total budget (Fig. 4.6). The remaining budget was allocated among other environmental agencies (see Table 4.2) with fewer responsibilities in protected areas.

Currently, managing protected areas requires more specialised and extensive knowledge than in the past, and the demands will be even greater in the future in the face of global changes. Strengthening capacity at the individual, institutional, and social levels is a priority (IUCN, 2005). Management models for the governance of protected areas, including by the government, must understand local and indigenous needs, priorities, practices, and values in order to meet the management objective or category of the protected area. Management models must establish strategies and

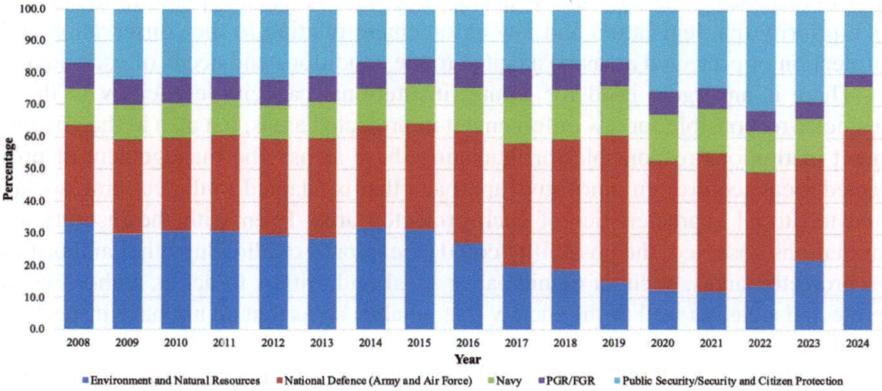

Fig. 4.5 Federal budget allocated to agencies by sector with functions in protected areas in Mexico in 2008–2024. **Source:** the authors with data Secretaría de Hacienda y Crédito Público (SHCP) (2025). *Until 2020, the federal criminal authority was called the Procuraduría General de la República (PGR) (Prosecutor General's Office). After 2020, its name changed to the Fiscalía General de la República (FGR) (Attorney General's Office)

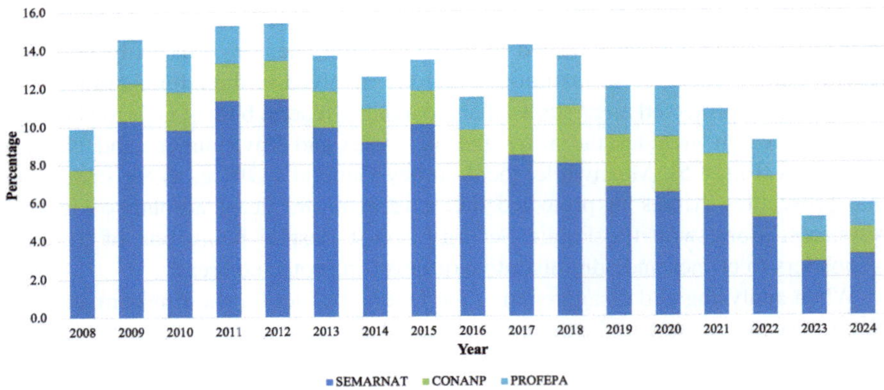

Fig. 4.6 Budget allocated to the main administrative institutions responsible for protected areas. **Source:** the authors with data from SHCP (2025)

policies that strike a balance between the costs and benefits of conservation and use (Valverde Garnica, 2015).

Conclusions

It is advisable to review and analyse the existing legal framework on protected areas at the national level in order to identify the responsible institutions and legal obligations, as well as to identify legal loopholes and provide information on the strengths and weaknesses of the law and its actual effectiveness when applied. In Mexico, it is a priority that legislation and law enforcement relating to the conservation and protection of protected areas be adapted to the conditions and needs of each area.

There is an urgent need for greater institutional capacity, especially in those agencies responsible for law enforcement in protected areas, but also in the government institutions responsible for managing these areas. The management of protected areas requires an innovative approach that is adapted to the cultural, social, and territorial characteristics of each protected area. Even with the institutional limitations observed, the possibility could be explored of allowing other authorities in protected areas, such as municipal or local authorities, to act as authorities in protected areas, as well as the society that inhabits areas that do not have inspection faculties.

In addition to CONANP personnel, local municipal authorities, or the community within or closest to protected areas must have authority, since without it, effective management is complicated, especially in remote areas where offices are far from protected areas and/or there are few personnel available to respond as quickly as possible to an emergency. In cases where it is not possible to grant surveillance authority to individuals or other institutions, working relationships should be developed and communication improved, particularly with law enforcement authorities.

References

Alers, M., Bovarnick, A., Boyle, T. J. B., Mackinnon, K., & Sobrevila, C. (2007). *Reducing threats to protected areas: Lessons from the field*. The World Bank. Available on http://documents.worldbank.org/curated/en/811641468157191779

Arriola-Padilla, V. J., Estrada-Martínez, E., Medellín-Jiménez, R., Gijón Hernández, A. R., Pichardo-Segura, L. A., Pérez-Miranda, R., & Ortega-Rubio, A. (2015). Áreas Naturales Protegidas del Centro de México: Degradación y Recomendaciones. In A. Ortega–Rubio, M. J. Pinkus-Rendón, & I. C. Espitia-Moreno (Eds.), *Las Áreas Naturales Protegidas y la Investigación Científica en México* (pp. 337–374). Centro de Investigaciones Biológicas del Noroeste S. C.. Universidad Autónoma de Yucatán and Universidad Michoacana de San Nicolás de Hidalgo. https://cobi.org.mx/wp-content/uploads/2016/01/2015_LIBRO-Las-%C3%A1reas-naturales-protegidas-y-la-investigaci%C3%B3n-cient%C3%ADfica-en-M%C3%A9xico.pdf

Balloffet, N. & Martin, A. (2007). Governance trends in protected areas: Experiences from the parks in peril program in Latin America and the Caribbean. The Nature Conservancy.

Bezaury-Creel, J., & Gutiérrez Carbonell, D. (2009). Áreas naturales protegidas y desarrollo social en México. In Comisión Nacional para el Conocimiento y Uso de la Biodiversidad (CONABIO) (Ed.), *Capital natural de México-Estado de conservación y tendencias de cambio* (pp. 385–431). Gobierno de México: CONABIO.

Bishop, K., Dudley, N., Phillips, A., & Stolton, S. (2004). *The uses and performance of the IUCN system of management categories for protected areas, IUCN: International Union for Conservation of nature*. Cardiff University: IUCN. Protected Areas Programme, UNEP-WCMC, WWF, Conservation International, Shell Corporation, British Petroleum (BP), International Council on Mining and Metals, UK. Available on https://portals.iucn.org/library/node/8520

Borrini-Feyerabend, G. (2007). *The IUCN protected area matrix. A tool towards effective protected areas systems*. IUCN.

Borrini-Feyerabend, G., Dudley, N., Jaeger, T., Lassen, B., Pathak Broome, N., Phillips, A., & Sandwith, T. (2014). *Governance of protected areas-from understanding to action. Best practice protected area guidelines series No. 20*. UICN. Available on https://portals.iucn.org/library/sites/library/files/documents/pag-020.pdf

Brenner, L. (2010). Gobernanza ambiental, actores sociales y conflictos en las Áreas Naturales Protegidas mexicanas. *Revista Mexicana de Sociología, 72*(2), 283–310. https://doi.org/10.22201/iis.01882503p.2010.002.17809

Carabias, J., de la Maza, J., & Cadena, R. (2003). *Capacidades necesarias para el manejo de áreas naturales protegidas*. América Latina y el Caribe. Available on https://iucn.org/es/content/capacidades-necesarias-para-el-manejo-de-areas-protegidas-america-latina-y-el-caribe

Castro-Salazar, J. I., & Camacho, M. O. (2020). La protección de Áreas Naturales Protegidas federales en México: estudio sobre la relación entre los "burócratas de a pie" y la superficie territorial de las ANP. *Sociedad y Ambiente, 22*, 22–47. https://doi.org/10.31840/sya.vi22.2073

Código Penal Federal. (2025). *Última Reforma Publicada en el Diario Oficial de la Federación el 16 de julio de 2025*. Cámara de Diputados del Honorable Congreso de la Unión. https://www.diputados.gob.mx/LeyesBiblio/pdf/CPF.pdf

Comisión Nacional de Áreas Naturales Protegidas (CONANP). (2016). *Plan de Acción para el Financiamiento Estratégico 2016–2018*. SEMARNAT. Available on https://www.gob.mx/conanp/documentos/plan-de-accion-para-el-financiamiento-estrategico-2016-2018

Comisión Nacional de Áreas Naturales Protegidas (CONANP). (2020). *Áreas naturales Protegidas*. SEMARNAT. Available on https://www.gob.mx/conanp/#1692

Comisión Nacional para el Conocimiento y Uso de la Biodiversidad (CONABIO). (2024). *¿Qué hacemos?* CONABIO. Available on https://www.gob.mx/conabio/que-hacemos

Consejo Nacional de Evaluación de la Política de Desarrollo Social (CONEVAL). (2014). *Diagnóstico del programa G005, Programa de Inspección y Vigilancia en materia de Medio*

Ambiente y Recursos Naturales. CONEVAL. Available on https://www.coneval.org.mx/Informes/Evaluacion/Diagnostico/Diagnostico_2014/Diagnostico_2014_SEMARNAT_G005.pdf

Correa, A. (2016). Gobernanza de las áreas protegidas y su necesaria incorporación en la ley que crea el Servicio de Biodiversidad. *Revista de Derecho Ambiental, 6*, 224–239. https://revistaderechoambiental.uchile.cl/index.php/RDA/article/view/43324

Costa Pellizzaro, P., Antunes Hardt, L. P., Hardt, C., Hardt, M., & Assef Sehli, D. (2015). Stewardship and management of protected natural areas: The international context. *Ambiente & Sociedade, 18*(1), 19–36. https://doi.org/10.1590/1809-4422ASOC509V1812015en

Day, J., Dudley, N., Hockings, M., Holmes, G., Laffoley, D., Stolton, S., & Wells, S. (2012). *Directrices para la Aplicación de las Categorías de Gestión de Áreas Protegidas de la UICN en Áreas Marinas Protegidas*. IUCN. https://portals.iucn.org/library/sitEs/library/files/documents/PAG-019-Es.pdf

de la Garza Buentello, H. A. (2014). *Elaboración de un diagnóstico de Operación en las Áreas Naturales Protegidas de carácter Federal para el Monitoreo, Control y seguimiento de acciones de Vigilancia, Prevención de incendios y Seguridad del personal Administrativo*. CONANP. Available on https://www.conanp.gob.mx/contenido/pdf/normatividad/estudios/DGOR/Diagnostico_de_op_en_ANP_monitoreo_control_y_seg_de_acciones_de_vig.pdf

de Montalvo Jääskeläinen, F. (2016). Potestad legislativa y evaluación ex post de las normas: hacia un mejor parlamento a través de la regulación inteligente. *Revista De Las Cortes Generales, 97-99*, 79–184. https://doi.org/10.33426/rcg/2016/97-99/18

Dearden, P., Bennett, M., & Johnston, J. (2005). Trends in global protected area governance, 1992–2002. *Environmental Management, 36*(1), 89–100. https://doi.org/10.1007/s00267-004-0131-9

Dudley, N. (2008). *Guidelines for applying protected area management categories*. IUCN. http://data.iucn.org/dbtw-wpd/edocs/PAPS-016.pdf

Fiscalía General de la República (FGR). (2024). Conoce la FGR. *Actuación y Misión*. Gobierno de México: FGR. https://fgr.org.mx/

García-Escudero, M. P. (2015). Regeneración del Parlamento. Transparencia y participación ciudadana. *Teoría y Realidad Constitucional, 36*, 171–216. https://doi.org/10.5944/trc.36.2015.16073

Gravez, V., Heylings, P., Rivadeneira, C., & Zambrano, N. (2011). *Gobernanza en las Áreas Protegidas Marinas y Costeras: el caso del Ecuador*. Fundación Futuro Latinoamericano. Available on https://www.ffla.net/wp-content/uploads/2021/03/FFLA_2011_Gobernanza-en-Areas-Protegidas-Marinas-y-Costeras_Ecuador.pdf

Gunningham, N. (2011). Enforcing environmental regulation. *Journal of Environmental Law, 23*(2), 169–2021. https://doi.org/10.1093/jel/eqr006

Henson, D., Malpas, R., & D'Udine, F. (2016). *Wildlife law enforcement in sub-Saharan African protected areas- a review of best practices*. IUCN. https://doi.org/10.2305/IUCN.CH.2016.SSC-OP.58.en

Hongjun, C., Haohao, S. & Xuesen, C. (2017). Analysis of environmental law enforcement mechanism based on economic principle. *IOP Conf. Series: Earth and Environmental* Science, 94, 012188. https://doi.org/10.1088/1755-1315/94/1/012188.

Johansson, V. (2010). Implementing chemical regulation: The role of inspectors. In J. Eriksson et al. (Eds.), *Regulating chemical risks: European and global challenges* (pp. 319–338). Springer Science+Business Media B.V.. https://doi.org/10.1007/978-90-481-9428-5_18

Kakira, L. M. (2010). *Monitoring law enforcement effort and illegal activity in selected protected areas: Implications for management and conservation*. Democratic Republic of Congo. Available on https://www.crgbc.center/prof-dr-leonard-kakira-mubalama/

Lausche, B. (2012). *Directrices para la legislación relativa a áreas protegidas*. IUCN. Available on https://portals.iucn.org/library/sites/library/files/documents/EPLP-081-Es.pdf

Ley de la Fiscalía General de la República. (2024). *Última Reforma Publicada en el Diario Oficial de la Federación el 1 de abril de 2024*. Cámara de Diputados del Honorable Congreso de la Unión. Availabe on https://www.ordenjuridico.gob.mx/leyes.php#gsc.tab=0

Ley de la Guardia Nacional. (2023). *Última Reforma Publicada en el Diario Oficial de la Federación el 16 de julio de 2023*. Cámara de Diputados del Honorable Congreso de la Unión. Availabe on https://www.ordenjuridico.gob.mx/leyes.php#gsc.tab=0

Ley General del Equilibrio Ecológico y Protección al Ambiente (LGEEPA). (2024). *Última Reforma Publicada en el Diario Oficial de la Federación el 1 de mayo de 2024*. Cámara de Diputados del Honorable Congreso de la Unión. Availabe on https://www.ordenjuridico.gob.mx/leyes.php#gsc.tab=0

Mackinnon, J., Mackinnon, K., Child, G., & Thorsell, J. (1986). *Managing protected areas in the tropics*. IUCN. Available on https://portals.iucn.org/library/sites/library/files/documents/1986-MacK-001.pdf

Mickwitz, P. (2003). A framework for evaluating environmental policy instruments: Context and key concepts. *Evaluation, 9*(4), 415–436. https://doi.org/10.1177/135638900300900404

Monk, J. (2012). *Reform of regulatory enforcement and inspections in OECD countries*. OECD Publishing. Available on https://coilink.org/20.500.12592/5cfxwg

Naughton-Treves, L., Buck Holland, M., & Brandon, K. (2005). The role of protected areas in conserving biodiversity and sustaining local livelihoods. *Annual Review of Environment and Resources, 30*, 219–252. https://doi.org/10.1146/annurev.energy.30.050504.164507

Nilsson, A. K. (2011). *Enforcing environmental responsibilities*. Uppsala University. https://www.diva-portal.org/smash/get/diva2:411546/FULLTEXT01.pdf

Organisation for Economic Co-operation and Development (OECD). (2004). *Assuring environmental compliance. A toolkit for building better environmental inspectorates in Eastern Europe, Caucasus, and Central Asia*. OECD Publishing.

Organisation for Economic Co-operation and Development (OECD). (2012). *La Evaluación de Leyes y Regulaciones: El Caso de la Cámara de Diputados de Chile*. OECD Publishing. https://doi.org/10.1787/9789264176362-es

Ortega-Rubio, A., Pinkus-Rendón, M. J., & Espitia-Moreno, I. C. (2015). *Las Áreas Naturales Protegidas y la Investigación Científica en México*. Centro de Investigaciones Biológicas del Noroeste S. C., Universidad Autónoma de Yucatán and Universidad Michoacana de San Nicolás de Hidalgo. Available on https://cibnor.repositorioinstitucional.mx/jspui/bitstream/1001/1026/1/2015_LIBRO-las-%C3%A1reas-naturales-protegidas-y-la--investigaci%C3%B3n-cient%C3%ADfi.pdf

Pautz, M. (2009). Trust between regulators and the regulated: A case study of environmental inspectors and facility personnel in Virginia. *Politics & Policy, 37*(5), 1047–1072. https://doi.org/10.1111/j.1747-1346.2009.00210.x

Pautz, M., & Schnitzer, M. (2008). Policymaking from below: The role of environmental inspectors and publics. *Administrative Theory & Praxis, 30*(4), 450–475. https://doi.org/10.1080/10841806.2008.11029665

Pliego-Moreno, V., Castillo-López, A., Castillo-López, J. E., Cuéllar, A., & Montiel-Bustos, R. (2016). *Diagnóstico del estado de la inspección y vigilancia en materia de pesca en el Alto Golfo de California*. Pronatura Noroeste, A. C. https://doi.org/10.13140/RG.2.2.26090.11205

Reglamento de la LGEEPA en materia de ANP. (2000). *Última Reforma Publicada en el Diario Oficial de la Federación el 30 de noviembre del 2000*. Cámara de Diputados del Honorable Congreso de la Unión. Gobierno de México. Availabe on https://www.ordenjuridico.gob.mx/leyes.php#gsc.tab=0

Reglamento Interior de la CONAGUA. (2012). *Última Reforma Publicada en el Diario Oficial de la Federación el 12 de octubre de 2012*. Cámara de Diputados del Honorable Congreso de la Unión. Availabe on https://www.ordenjuridico.gob.mx/Documentos/Federal/pdf/wo88745.pdf

Reglamento Interior de la SEMARNAT. (2025). *Última Reforma Publicada en el Diario Oficial de la Federación el 14 de marzo de 2025*. Cámara de Diputados del Honorable Congreso de la

Unión. Availabe on https://www.dof.gob.mx/nota_detalle.php?codigo=5751954&fecha=14/03/2025#gsc.tab=0

Rivas Toledo, A. (2006). *Gobernanza de los Sistemas Nacionales de Áreas Protegidas en los Andes Tropicales: Diagnóstico regional y análisis comparativo*. IUCN. Available on https://portals.iucn.org/library/sites/library/files/documents/2006-065.pdf

Rodríguez-Quiroz, G., González-Ocampo, H. A., Barba, E., Jiménez-Badillo, L., Pinkus-Rendón, M. J., Pinkus-Rendón, M., & Ortega-Rubio, A. (2015). Aspectos socioeconómicos de la pesca artesanal en las áreas naturales protegidas. In A. Ortega–Rubio, M. J. Pinkus-Rendón, & I. C. Espitia-Moreno (Eds.), *Las Áreas Naturales Protegidas y la Investigación Científica en México* (pp. 195–210). Centro de Investigaciones Biológicas del Noroeste S. C., Universidad Autónoma de Yucatán and Universidad Michoacana de San Nicolás de Hidalgo. https://cibnor.repositorioinstitucional.mx/jspui/bitstream/1001/1123/1/PUB-CAPITULOS-LIBROS-974.PDF

Rose, G. L. (2011). *Gaps in the implementation of environmental law at the national, regional and global level*. UNEP. Available on https://ro.uow.edu.au/articles/report/Gaps_in_the_Implementation_of_Environmental_Law_at_the_National_Regional_and_Global_Level/27801891?file=50574411

Secretaría de Hacienda y Crédito Público (SHCP). (2025). *Transparencia presupuestaria. Observatorio del gobierno*. SHCP. Available on https://www.transparenciapresupuestaria.gob.mx/

Secretaría de Relaciones Exteriores (SRE). (2025). *Buscador de tratados vigentes*. SRE. Available on https://cja.sre.gob.mx/tratadosmexico/buscador

Secretaría del Convenio sobre la Diversidad Biológica. (2004). *Programa de Trabajo sobre Áreas Protegidas*. Secretaría del Convenio sobre la Diversidad Biológica. https://www.cbd.int/doc/publications/pa-text-es.pdf

Simonian, L. (1999). *La defensa de la tierra del jaguar: Una historia de la conservación en México*. SEMARNAP-Instituto Nacional de Ecología. Available on http://www.conabio.gob.mx/institucion/proyectos/resultados/FichapubT007.pdf

Stab, S. & Henle, K. (2009). Research, management, and monitoring in protected areas. In F. Gherardi, C. Corti & M. Gualtieri. Biodiversity conservation and habitat management (pp. 128–152). EOLSS Publications. https://www.eolss.net/sample-chapters/c12/E1-67-01-03.pdf

Tremper, C., Thomas, S., & Wagenaar, A. (2010). Measuring law for evaluation research. *Evaluation Review, 34*(3), 242–266. https://doi.org/10.1177/0193841X10370018

International Union for Conservation of Nature (IUCN). (2005). *Beneficios más allá de las fronteras. Actas del V Congreso Mundial de parques de la UICN*. IUCN. Available on https://redeuroparc.org/wp-content/uploads/2022/02/durban_actas_durban.pdf

Valverde Garnica, Á. (2015). Gobernanza ambiental en áreas protegidas. El caso del Parque Nacional y Área Natural de Manejo Integrado Cotapata. *Revista Integra Educativa, 8*(3), 71–85. http://www.scielo.org.bo/scielo.php?pid=S1997-40432015000300006&script=sci_abstract

Wyatt, T., van Uhm, D., & Nurse, A. (2020). Differentiating criminal networks in the illegal wildlife trade: Organized, corporate and disorganized crime. *Trends in Organized Crime, 23*, 350–366. https://doi.org/10.1007/s12117-020-09385-9

Zielińska, A. (2010). Management system of protected areas and its monitoring in Poland. *Economics & Sociology, 3*(2), 121–127. https://doi.org/10.14254/2071-789x.2010/3-2/13

Wyatt, T. (2015). Mapping the links between conflict and illegal logging. In A. Brisman, N. South, & R. White (Eds.), *Environmental crime and social conflict: Contemporary and emerging issues* (pp. 41–56). Routledge. https://doi.org/10.4324/9781315580012

Part II
The Mexican War on Drugs and Protected Areas

Overview

This section addresses the context of the emergence of drug trafficking in Mexico and explores how decades of "peace pacts" (*pax narcotica*) were made between criminal groups and government entities and how this facilitated the consolidation of the "Major Drug Cartels".

This context is analysed from the perspective of the protected areas due to their importance as strategic territories for the production, processing, and transit of drugs and is analysed temporally: before, during, and after the war (post-war period); using the experiences and testimonies of environmental inspectors and directors as public servants of the protected areas a support and guiding thread.

Chapter 5 explores through a historical-documentary analysis the emergence of drug trafficking in Mexico, as well as its consolidation into large drug cartels and its expansion and territorial settlement throughout the country, including territories dedicated to environmental conservation, rural areas, and indigenous communities. It also considers temporally from the emergence and consolidation of drug trafficking until the first half of the 1980s.

Chapter 6 explores the alliances and conflicts between criminal organisations and government actors, emphasising how these relationships facilitated the exploitation of protected areas for drug trafficking activities. This chapter continues the diachronic analysis of drug trafficking in Mexico.

Chapter 7 analyses the ways in which environmental conservation activities were conducted in protected areas, considering the context of the pax narcotica since the 1990s, by documenting how environmental conservation activities were affected by the presence of drug traffickers in protected areas as well as the ways in which inspectors and directors adapted to continue conservation activities within protected areas.

Chapter 8 explores how the Mexican War on Drugs modified the dynamics of environmental conservation within protected areas, including the development of personal protection strategies for the personnel of the National Commission of Natural Protected Areas (CONANP), researchers and students, rural and indigenous communities, as well as the development of public policies aimed at strengthening security within these territories.

Chapter 5
The Rise of Drug Trafficking in Mexico

Introduction

Records of human consumption of (natural) drugs can be traced back to before 5700 BC in regions of the Mediterranean and Asia (Mack & Joy, 2000). However, in the last two centuries, drugs used for medicinal, recreational, and religious purposes have been subject to regulation and/or prohibition. At the end of the nineteenth century and the beginning of the twentieth century, drug use was legal worldwide. Products such as "Bayer Heroin Syrup" produced by Bayer for respiratory diseases and symptoms, coca wine first produced in Bolivia as "Elixir de Coca" and later produced in France as "Mariani wine" for singers and athletes, marijuana cigarettes marketed by Grimault & Co. for asthma, coughs, insomnia, and rheumatic pains (Bonnie & Whitebread, 1999; Mack & Joy, 2000), as well as medicinal treatments that were supplemented with cocaine pills and marijuana cigarettes (Musto, 1997).

The abuse of opiates and marijuana soon became a public health problem in many countries, with the United States being the main proponent of drug prohibition. However, prior to American attempts to ban the drug, other countries began to regulate its consumption and production. For example, in 1810 during the Qing Dynasty, a decree was issued prohibiting the sale of opium gum under penalty of death (Grillo, 2011; Valdés, 2013), while in Mexico, the first efforts were made to ban "fumaderos" (places where opium was consumed) in the north-west of the country. By 1908, the Opium Commissioner in the United States argued that drug use was a serious problem that required "protecting Americans from drugs coming from outside the country". Consequently, rhetoric was used claiming that "caused Chinese opium users to rape white women and black men on cocaine to acquire superhuman strength with which to murder white men" (Astorga, 2005, p. 51), thus

Supplementary Information The online version contains supplementary material available at https://doi.org/10.1007/978-3-032-18218-0_5.

J. L. Carpio-Domínguez et al., *Drug Trafficking and Protected Areas in Mexico*,
https://doi.org/10.1007/978-3-032-18218-0_5

representing an external threat and making the United States a leader in the global campaign against drug production and consumption.

Subsequently, in 1909, the first meeting was held in Shanghai, China, to establish controls on the production and consumption of drugs, mainly opium and its derivatives. Three years later, in January 1912, the International Opium Convention was held in The Hague, at which 12 countries ratified strategies for the control of opium and its derivatives, which became effective until June 1919 (United Nations, 1912). These efforts to regulate drug production and consumption laid the groundwork for Mexico in subsequent years to: (a) implement stricter controls on drug production and consumption within its national territory and (b) restructure the entire drug production and consumption system in the country, leading to prohibition and the emergence of criminal groups dedicated to drug trafficking.

For the diachronic analysis of the emergence of drug trafficking in Mexico in this chapter, we will draw on the works of Astorga (2005), Grillo (2011), and Valdés (2013) due to their important contribution to recovering historical information on this phenomenon and their status as the leading exponents of the historiography of drug trafficking in Mexico. In this chapter, we do not intend to conduct an in-depth historical analysis, but rather to highlight the relevant events that lay the foundations for the emergence of drug trafficking in the country diachronically with the decrees on environmental conservation areas during this historical period.

From the Origin to Drug Prohibition in Mexico

The origins of drug trafficking in Mexico can be traced back to the late nineteenth century and the first two decades of the twentieth century, until its prohibition in 1926. However, during that period in Mexico, the production, consumption, and importation of drugs were legal. During the Porfiriato, the importation of opium into Mexico ranged from 800 kg to 12 tonnes between 1888 and 1911, with an average of 1 tonne per year, coming from the United States, Europe, and Asia. According to Astorga (2005), these imports were intended to supply products for human consumption that were legally sold in pharmacies. Cordial wines, coca wines, and marijuana cigarettes were offered in pharmacies throughout the country.

Historical records show that the first poppy seeds to enter Mexico were brought by people of Chinese origin during the construction of the railway and for mining work, first to the state of Baja California and then to Sinaloa. In addition to bringing the seeds, they knew how to sow and harvest them, as well as how to convert the flowers into opium for personal consumption. Therefore, most opium smokers in those years were found in cities in these states.

In cities such as Culiacan, Mazatlán, and Navolato in the state of Sinaloa, Mexicali in Baja California, Guadalajara in Jalisco, Hermosillo in Sonora, Chihuahua, Tamaulipas, and Mexico City (formerly known as the Federal District), "fumaderos" or smoking dens were places frequented first by people of Chinese origin and later by Mexicans who became frequent customers in the consumption of opium (Valdés, 2013). With complaints from citizens and Mexico's ratification of the Shanghai International Opium Convention in May 1912, policies restricting opium were established in the country.

With the armed uprising of the Mexican Revolution (1910–1917), national efforts to regulate the production and consumption of opium and marijuana were affected. In northern Mexican states, the Revolution left an economic crisis and populations destroyed and were in poverty. Among the strategies to address these crises, for example, in Baja California, the production and sale of liquor, brothels and casinos were taxed, but so was the opium trade, even despite recent national and international regulations (Capó-Valdivia, 2015).

However, despite restrictions on opium, international networks aimed at bringing opium to the American market could already be identified in northwestern Mexico. Various studies have documented the presence of international networks in the northwestern region of the country (Campos, 2010; Capó-Valdivia, 2015; González-Félix, 2002; Pérez Montfort, 1997). For example, it was documented that raw opium was probably grown in China or India and exported to the United States via transatlantic ships departing from Macao, Hong Kong, and Singapore. These ships arrived directly in San Francisco, California, with the aim of entering Mexico via the border to be processed and returned to the United States. In addition, it has been documented that raw opium also entered Mexico via ships that anchored in ports such as Salina Cruz, Manzanillo, Guaymas, Mazatlán, and Ensenada, coming from New York, England, Italy, Germany, and Japan (Astorga, 2005; Paoli-Bolio, 2008) (Fig. 5.1).

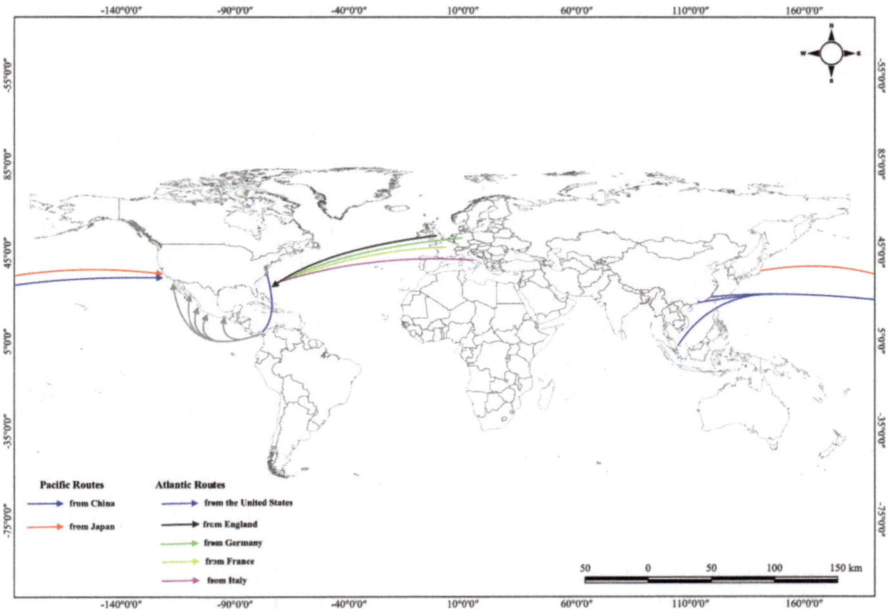

Fig. 5.1 Opium imports to Mexico in the late nineteenth century. *Source*: the authors with data from Astorga (2005), Paoli-Bolio (2008), and Valdés (2013)

Once raw opium entered Mexico, it was transported inland by land using horses, donkeys, mules, and trains. It was then processed in factories or refineries in the north-west of the country, turning it into smokable opium and other derivatives such as morphine and heroin. A small portion remained in the region, but most was packaged and illegally re-exported to the state of California in the United States (Capó-Valdivia, 2015; Sandos, 1984). These activities were mainly carried out by people of Chinese origin during the 1910s and 1920s, who benefited financially from this illegal market and strengthened their networks and their families' economies (González-Félix, 2002).

It should be highlighted that during this period, Mexico's first Protected Natural Area, named "Desierto de los Leones National Park", was created, transforming land expropriated by President Sebastián Lerdo de Tejada in 1876 into a national park in 1917, with the aim of protecting springs and providing water to Mexico City in response to the increase in deforestation in the area (De la Maza, 1999). Historical records have documented the existence of marijuana plantations in forested areas of central Mexico (Astorga, 2005); however, no records of plantations were found in this first protected natural area.

On the other hand, marijuana as a natural drug has no documented exact date when it began to be produced in Mexico; however, its recreational use did not represent a public health problem until 1920. The main characteristic of marijuana for recreational purposes was its low production cost, and prior to its prohibition, its production supplied the local market in Mexico and the United States. In Mexico, seizures of marijuana and crops were documented at that time in states such as Sinaloa, Durango, Mexico City, Puebla, San Luis Potosí, Oaxaca, Guerrero, Querétaro, Guanajuato, Morelos, Hidalgo, Tlaxcala, Sonora, Colima, Veracruz, Nuevo León, Coahuila, Tamaulipas, and the State of Mexico, representing 18 (56.3%) of the country's 32 states (Astorga, 2005; Valdés, 2013).

Although opium had been regulated in the United States since 1914, Mexico did not enact its own legal regulations until 1920, when it issued the first legal provisions on what was termed "the cultivation and trade of products that degenerate the race", partially prohibiting the cultivation and trade of marijuana (Grillo, 2011). Meanwhile, the poppy, also known as opium poppy, and its derivatives were regulated until 1925 through a decree by President Plutarco Elías Calles, which established criteria for the importation of opium, including morphine and cocaine. By 1926, the prohibition of marijuana was followed by legislation against opium and its derivatives.

It should be noted that during this period, efforts to protect natural areas that provided ecosystem services in various parts of the country were consolidated during the presidential terms of Álvaro Obregón (1920–1924) and Plutarco Elías Calles (1924–1928). Among the most representative are the declaration by presidential decree of the Guadalupe Island Forest reserve in Baja California in 1922, the Sierra de Guadalupe north of Mexico City in 1923, the national lands of Izatán in Jalisco in 1924, while in 1924, various forest areas in Puebla and the State of Mexico were declared protected (De la Maza, 1999).

According to Valdés (2013), the regulatory differences between the United States and Mexico regarding drugs became an opportunity for those who produced and processed these drugs. This gap in prohibition led to higher drug prices in the United States, making it a potentially exploitable and profitable market for producers in Mexico.

This phenomenon had previously been documented in the Pacific and Indian Oceans in the nineteenth century during the Qing Emperor's opium ban in China. According to Grillo (2011), "this ban [and its potential profitability] created the first smugglers in the form of friendly and elegant pipe-smoking gentlemen of the British Empire" (p. 48); the East India Company smuggled thousands of tonnes of opium from India to China in what can be considered, despite Chinese efforts to ban trafficking through the confiscation of English ships carrying opium, "the East India Company represented the first drug cartel and the British Navy the first armed and violent arm of illegal drug trafficking" in the world (Grillo, 2011, p. 48).

This discrepancy in drug prohibition policies allowed Chinese producers in Mexico to supply the domestic market and export to the United States, consolidating their position in profitable markets due to the rise in costs following prohibition and other factors such as high demand and the social status associated with drug use in smoking dens or opium addiction, which was considered an elegant, expensive, sumptuous, aristocratic vice through the consumption of morphine (known as morphinism), characterised by its association with the upper and middle classes.

The prohibition of drugs in Mexico took place in 1926, representing a radical change in the ways in which drugs were produced, transported, and sold within the country and exported, although previously, in 1925, President Plutarco Elías Calles had requested a bilateral treaty from the United States government to combat drug smuggling, ordering Mexican authorities to increase actions against traffickers and users of opium, heroin, and cocaine, including weapons and alcohol (Enciso, 2010; Valdés, 2013).

Strengthening of Drug Production in Mexico

Following the ban in Mexico, in the 1930s, there were records of morphine and opium consumption centres in cities in the west and north of the country, such as Manzanillo in the state of Colima, Guadalajara in Jalisco, Mexicali in Baja California, Ciudad Juárez in Chihuahua, Culiacan in Sinaloa, and Hermosillo in Sonora (Valdés, 2013). In terms of the profitability of the business, by the end of the decade, profits of seven million Mexican pesos per year (approximately $350,000 US dollars) could be accounted for from the purchase and distribution of just 1 kg of heroin, which cost 10,000 Mexican pesos (approximately US$500 today) (Astorga, 2005, p. 40).

These settlements of Chinese communities supplied the domestic and American markets, not only through drug production within the country, but also through imports. Since the 1920s, Chinese residents settled in Baja California, Durango,

Sonora, and Sinaloa had land and farms, labour, and knowledge to produce poppies and opium. For example, in the early 1920s, there were Chinese communities in Navolato and Culiacán in the state of Sinaloa, and by the following decade, there were communities in other municipalities in the state, and poppy fields were discovered in Sonora, Jalisco, Michoacán, Guanajuato, and Xochimilco in Mexico City (Valdés, 2013).

In addition to domestic production, there was also illegal opium imported by sea. Astorga (2005) documented that at the end of 1930, 516 kilograms of opium were seized in Manzanillo in the state of Colima, on the Pacific coast, so it is reasonable to assume that there must have been trafficking networks to establish communication and collaboration networks between Chinese people in Mexico and China. With a Mexican government that did not exercise much control, the Chinese in Mexico traded with other Chinese in the United States. Even so, Grillo (2011) points out that just as the British challenged China's opium ban, the Chinese now challenged the Mexican and American bans.

Mexico's northern border was (and continues to be) an ideal location for illegal trafficking. The characteristics of the border, with its large cities such as Ciudad Juárez and Tijuana and other medium-sized cities such as Mexicali and Nogales, meant that it had strong links with southern regions where drugs were produced, such as the Sierra Madre Occidental, mainly in the states of Chihuahua, Sonora, Sinaloa, and Durango (Grillo, 2011; Valdés, 2013). However, cities such as Nuevo Laredo, Reynosa, and Matamoros, which are the southernmost cities on the border and close to central Mexico, also played a role (Correa-Cabrera, 2016). According to Grillo (2011) "everything from Aztec ceremonial skeletons to Browning machine guns and even white tigers" (p. 53) has passed through this border, demonstrating the historical importance of Mexico's northern border in various forms of illegal trafficking, including wildlife trafficking, which is less well documented but has continued to this day (e.g. Arroyo-Quiroz & Wyatt, 2019; Carpio-Domínguez, 2021; Carpio-Domínguez et al., 2018, 2022, 2023).

During this decade, the opium and marijuana trade became established as illegal activities based on local networks mainly involving communal landowners and rural communities, with three main factors (Valdés, 2013):

1. Government neglect of rural communities and those far from urban centres.
2. Social self-protection against the authorities through warnings when there were drug raids, for example, by ringing the local church bells and.
3. Although drug production appeared to be individualised, protection and marketing were organised collectively. These factors were reinforced by feelings of solidarity among community members, or by family, friendship, or economic ties.

In cities and semi-urban settlements, the situation was different. At the end of the 1930s, several figures linked to the drug trade in cities in the centre of the country stood out. According to historical records, although there was no large-scale criminal association in the Mexican marijuana market, or anything that could be called a national drug cartel, some of the groups that had formed had prominent representatives. One example was Felisa Velázquez, known as the "Queen of Marijuana" who owned the marijuana plantations in Cholula, Puebla. She established a marijuana consumption network in a very profitable consumption centre, so much so that when she was arrested, the prisoners in Lecumberri (the largest prison in Mexico at the time) rioted over the lack of drugs inside the prison, but this network also supplied military barracks, which were supplied with drugs by several women, leading to the suspicion that this marijuana trafficking network was made up of several female drug dealers (Valdés, 2013).

These networks also distributed drugs (mainly marijuana) among the wealthy and upper classes in Mexico City. Another woman known as "Lola La Chata" who lived in a luxurious residence in Lomas de Chapultepec, one of the wealthiest neighbourhoods in Mexico City, distributed marijuana to wealthy people in the city. Sixto Vargas, known as "El Tigre del Pedregal" also stands out, belonging to another of the city's wealthiest neighbourhoods.

Meanwhile, Félix Sánchez, known as "El Zar de las Drogas" (The Drug Czar), together with his brother Othón Sánchez, were leaders of one of the largest gangs in Mexico City and were involved in the sale of opium and marijuana in one part of the city. If these historical records are accurate, these were local marijuana and opium trafficking networks (Valdés, 2013).

This overview of the emergence of drug trafficking in Mexico highlights two important points: (1) those who led the drug trade in the city accumulated wealth, living in the wealthiest areas of the cities and (2) there are no historical records of the quality of life of farmers and drug producers in the countryside. However, Valdés (2013) infers that this population did not receive fair payments for the purchase of drugs compared to the leaders in the city. This is important because drug production (planting, harvesting, and packaging) took place in areas far from urban settlements and, consequently, far from the health authorities responsible for drug control.

In the special case of marijuana produced in the centre and west of the country, it was particularly difficult to distribute it to the north, mainly to the United States. This demand was met by Chinese producers in the north of the country, in states such as Sonora, Sinaloa, and Chihuahua, due to their proximity to the border and the infrastructure of relationships they had in place to achieve this, and they turned to small Mexican producers to supply the demand. Although there are no historical records to prove this, Valdés (2013) infers that this may have been the case until Mexican producers began to establish their own networks both nationally and in the United States.

Regarding the protection of natural areas, in the early 1930s, President Pascual Ortiz Rubio declared the protection of forest areas in the Mexico Valley watershed, the forests of the Tocuila region in Veracruz, properties on the edges of Cerro del Vigia in Mazatlán, Sinaloa, and promoted reforestation efforts in areas of the State of Mexico and Puebla. By 1934, President Abelardo Rodríguez established guidelines for the protection of forest areas that supplied water for irrigation and human consumption in urban settlements, such as the forests of Cañada de Contreras in Mexico City, the San Elías property in Chihuahua, and land on the periphery of Guadalajara in Jalisco, and to protect against landslides, forest areas near the city of Río Blanco in Veracruz and the lands of the La Carbonera and Río Blanco River basins were protected (De la Maza, 1999). However, no records have been found on drug production during this time in these environmental conservation areas, although it cannot be denied.

Mexican Emancipation from the Chinese Drug Market

The emancipation of drug production by Mexican producers from Chinese trade networks can be traced back to a strategic racist campaign of "anti-Chinese" sentiment, mainly in northwestern Mexico. Although it had been brewing for several years, it became most evident during the 1930s. Historical records recovered by Grillo (2011) show that slander against people of Chinese origin by Mexicans, driven by politicians and criminals.

The American consul in Ensenada, Baja California, documented in a report sent to Washington the growing anti-Chinese sentiment in the region, highlighting that a smuggler from the states of Baja California, Sonora, and Sinaloa, whose surname was Segovia, was financing violent anti-Chinese groups with the aim of taking over Chinese opium production businesses.

Furthermore, these records highlight that racial tension towards Chinese people intensified in the streets of northwestern Mexico, which soon escalated into lynchings. Grillo (2011) mentions that a student named Manuel Lazcano, originally from Sinaloa, joined the lynch mobs at that time and later became an important figure in the political and judicial spheres, serving as the Attorney General of Sinaloa for three 6-year terms. Although he later repented of his participation, he left a record of the events of this anti-Chinese crusade in the state of Sinaloa.

According to Grillo (2011), Lazcano's records left evidence of the cruelty of these activities. Mobs roamed the streets searching for and hunting down Chinese people. Once they had detained them, they tied their hands and feet and imprisoned them in clandestine jails, walled compounds, or abandoned houses while they waited to gather a larger group of people. They were then loaded onto train carriages and expelled from the state. Subsequently, the Chinese people's properties and houses were appropriated by Mexicans.

In other places such as Ciudad Juárez, gangsters did not bother to detain and expel the Chinese; they simply murdered them. Among the documented cases was a gunman named "El Veracruz" who cornered and murdered 11 Chinese men working in the opium market. It was documented that this gunman's boss was a woman named Ignacia Jasso, known as "La Nacha". According to Grillo (2011), Lazcano himself acknowledged that this "ethnic cleansing" due to drug trafficking in northwestern Mexico was similar to the activities that the German people were carrying out against the Jews during those same years. Grillo (2011, p. 59) quotes the following passage from Lazcano's words:

> We have seen in films the brutal repression suffered by the Jews [...] scenes of how they were transported like animals. Well, the same thing happened in Sinaloa, but with the Chinese. Seeing those scenes in real life was shocking.

The emancipation of drug production and trafficking in Mexico in the late 1930s reflects the strategy and use of social, political, economic, and cultural resources that centred on violence in all its forms and which, historically, has characterised drug trafficking, not only in Mexico, but also in other countries where criminal associations have been documented for various forms of illegal trafficking, such as Italy, the United States, Japan, China, Colombia, etc.

Monopolisation of the Mexican Drug Market

Once the Chinese population was eliminated from opium production in Mexico in the 1930s, the following decade was characterised by the consolidation of the Mexican drug market. This consolidation was contextualised by the rise of the Second World War, as the pharmaceutical industries that supplied raw materials to produce morphine and other opium-derived medicines were located in countries at war where transport routes were unsafe, such as Turkey, Afghanistan, China, and India.

Although there are rumours of an unofficial agreement between the US government and Mexican authorities to supply the demand for morphine in the context of the war, there are no records to support this. However, what is clear is that many American soldiers became addicted to morphine and heroin during the war and continued their addiction upon their return to the United States. This increase in drug users in the United States showed that the drug production business in Mexico was responding to demand.

According to Valdés (2013) and Astorga (2005), the United States Treasury Department estimated that Mexico produced 60 tonnes of opium in 1943 alone, which was three times more than the previous year. In this context, by 1944, the record for drug production areas in Mexico was held by a region called Metates in the state of Durango in the Sierra Madre Occidental, with a total of 230 hectares of poppy cultivation.

However, these hectares in the Metates region of Durango were destroyed in an operation by the Attorney General's Office, adding to a series of operations in Mexican territory, mainly in the area known as the "Triángulo Dorado" (Golden Triangle) which corresponds to municipalities within the Sierra Madre Occidental in the states of Sinaloa, Chihuahua, and Durango (Figs. 5.2 and 5.3) (Rodríguez, 2022). The Golden Triangle refers to a geographical region which, due to its location and climatic and orographic characteristics, facilitates the production of drugs (poppy and marijuana).

It should be noted that the state of Sinaloa had the highest number of municipalities with drug plantations, among which the municipalities of Badiraguato, Culiacán, Sinaloa de Leyva, Mocorito, Cosalá, Choix, El Fuerte, and San Ignacio are highlighted. Badiraguato being the most representative in terms of both production quantity and poppy quality. According to Astorga (2005), Grillo (2011), and Valdés

Fig. 5.2 Pine-oak ecosystem in the mountainous areas of the Golden Triangle in northwestern Mexico. Photograph: José Luis Carpio Domínguez

Fig. 5.3 Geographical location of the Golden Triangle in northern Mexico. *Source*: the authors. *Note*: There are no records mentioning each drug-producing municipality in the Golden Triangle in this region of the Sierra Madre Occidental. This map only shows those mentioned by Astorga (2005), Grillo (2011), and Valdés (2013)

(2013), the poppies produced in this municipality were of even better quality than those from Smyrna in Turkey.

The demand for opium gum led to an increase in poppy production in northern Mexico, particularly in the Golden Triangle, during this decade, allowing for the consolidation of a drug production structure even larger than when the Chinese led the market. According to historical records, Mexican government efforts against drugs had little impact. The US government even financed anti-drug campaigns in northern Mexico during the presidency of Manuel Ávila Camacho (1940–1946) and requested increased funding from the United States to eradicate 25% of the poppy-growing area in Mexican territory.

According to Astorga (2005), in 1948, there were more than 10,000 poppy fields in Mexico, producing more than half of the opium, morphine, and heroin that was illegally imported into the United States. Although there are no historical records of the formation or consolidation of drug trafficking as we know it today during the 1940s and 1950s, it was undoubtedly during this period that the structure of drug trafficking began to take shape for the next decades.

Collaboration Between Drug Producers and the Judicial and Political System

The main characteristics of drug production during the 1940s were the alliances between drug producers and their leaders with high-ranking police officers and politicians. For example, in 1947, there were scandals regarding the relationship and links between the governors of the states of Baja California Norte, Sonora, and Sinaloa and the drug trade. It was even mentioned in national newspapers that the military and then-governors of Sonora and Sinaloa led drug trafficking gangs and owned planes and airstrips to transport opium. At that time, it was documented that opium was trafficked by air through a structure and infrastructure of private and public airports, as well as airstrips in the Sierra Madre Occidental.

Astorga (2005) even cites a publication from the newspaper *El Universal* that documented the existence of international gangs composed mainly of Americans, who financed poppy producers and managed production with politicians such as municipal presidents and directors and former directors of the judicial police, mainly in the state of Sinaloa. The public servants' role was to organise the collection of fees owed by the farmers who owned the land where the poppies were grown, as well as the salaries of those responsible for processing the drug in laboratories.

These relationships between financiers, producers, businesspeople, politicians, and public servants in the police and judicial system in drug production led to protectionism characterised by impunity and evasion of justice. Furthermore, this situation was exacerbated by "social and political" protectionism, in which people and the police knew who produced drugs, who sold them, who transported them, and who protected them, but no one reported them or took legal action (Astorga, 2005; Valdés, 2013).

This situation must be understood at the local level, in a socio-historical context in which drug production was a subsistence activity in communities far removed from government management of basic services, employment opportunities, education, health services, and profitable sources of income. Nevertheless, the production and sale of opium provided the economic means to meet basic needs and even leisure activities that were not provided for by the Mexican government.

Regarding (natural) protected areas, in the late 1940s, several National Forest Reserves were declared, such as the Sierra de Juárez, and the mountains that form the Sierra de San Pedro Mártir were declared a National Park, both in Baja California Norte in northeastern Mexico (De la Maza, 1999). In the central region of the country, the continuous territorial and demographic growth of Mexico City led to the Valley of Mexico basin being declared a forest protection zone, overlapping previous decrees on this area, which were never enforced and were repealed or modified. In the early 1950s, the Río Florido basin in the state of Chihuahua was also declared a forest protection zone, and in late 1950s, the Lagunas de Montebello in the state of Chiapas was declared a National Park (De la Maza, 1999).

Opium production in the late 1940s and early 1950s was characterised by the consolidation of trafficking networks along various routes to the US border, including those not previously used. Traditionally, land routes were mainly used, primarily by train or pack animals, as well as sea routes using small boats and motorboats. However, air routes had not been exploited except in isolated cases in the late 1940s. This situation changed during the 1950s due to the potential benefits of transporting opium by light aircraft from the production sites in the Golden Triangle in the Sierra Madre Occidental to the border. This route presented fewer risks of theft, arrest, or loss of the product, as well as reducing transport times (Valdés, 2013).

In the 1950s, the consolidation of Mexican drug production networks was characterised by the scattered participation of farmers and small landowners in thousands of plantations, which in one way or another limited the Mexican government's efforts to end production, mainly of opium. Production was consolidated through small agricultural bourgeoisies with strong ties to smaller producers, police, municipal presidents, and other government entities. During this time, the first laboratories for processing opium gum into heroin or morphine were established, giving the business a boost towards more industrialised production.

Furthermore, it should be noted that during this period, there was a constant American presence in drug-producing territories, both as buyers of opium and as financiers of production. This shows that the consolidation of production networks had grown beyond local production. This showed that the US investment in drug production in the Golden Triangle, well documented by Astorga (2005), Grillo (2011), and Valdés (2013), could be considered a commercial and financial partnership between large binational companies. It was no longer a matter of networks between Chinese relatives and families, but rather larger criminal organisations with strong links between the United States and Mexico.

The involvement of police officers, mayors, and other public servants who provided protection to producers and leaders of the nascent Mexican drug trafficking industry allowed public officials to amass great wealth. This enrichment came mainly from a system of taxes on small producers, informal agreements, payments for favours, and the sale of protection, as was the case with the Sicilian mafia. The relationships of these public servants were not only composed of links with opium producers, but also with high-level government officials, including governors, deputies, senators, and even leaders of the fight against drug trafficking from the federal government.

According to Valdés (2013), these criminal associations should not be considered to have nullified the rule of law during the emergence of drug trafficking in Mexico, mainly because drug trafficking was organised and directed by part of the state itself, through strategic public servants and the organisational structure of government institutions. This prevented the formation of external criminal groups that offered protection to other private groups. In this sense, the state itself coordinated and directed drug production and provided the necessary protection; therefore, "a part of the state nullified the state in its entirety" (Valdés, 2013, p. 102).

With this new route for opium trafficking, in 1953, the Civil Aviation Department of Mexico's Ministry of Communications and Public Works decreed the suspension of commercial flights at airfields in the states of Sonora, Sinaloa, Chihuahua, and Durango, as well as the cancellation of the Culiacan Aviation School in Sinaloa due to suspicions that the school was being used for opium transport. Astorga (2005, p. 71) notes that during the cancellation of the Culiacán Aviation School, the owner and flight instructor was accused of trafficking many kilos of opium gum with the following testimony:

> I went to complain and told them that I hadn't transported kilos of opium gum, but tons [...] I asked them why they didn't stop Aeronaves de México and the railroad, which had transported more gum than my planes. A few days later, they let me fly again.

During this decade, the illegal drug economy became increasingly complex in terms of both quantity and structure, mainly because the market was composed of many parts that performed specific functions for both production and export. This meant thousands of independent producers, well-organised marketing networks using air, land, and sea transport, US financing systems, commercial alliances between producers in both countries, and the participation of a diverse range of social actors such as peasants, farmers, professionals, students, families, politicians, police, and military personnel.

To understand the socio-historical context, an editorial from a Culiacan newspaper by the Attorney General of Sinaloa (1952–1952) is cited, this cite was recovered by Astorga (2005, pp. 88–89), which highlights the social and political reasoning behind drug production in the state of Sinaloa:

> If we were morphine addicts or heroin users, or simply enjoyed smoking opium, we would not have to defend, as we do and have been doing, the cultivation of opium poppies, which is so tenaciously and systematically combated by our government in order to satisfy foreign needs or to fulfil international commitments made without prior and thorough study and without the slightest defence of national interests.

Therefore, the apology for drug production, under the discourse of a ban imposed by the Mexican government with foreign interests and without considering the economic benefit to the local and national context, strengthened and validated this fledgling business in public opinion. According to Grillo (2011), during this time, opium production was so popular that the Culiacan baseball team was called "Los Gomeros" (The Gum Makers), referring to the production of opium gum as a source of economic income for society. It was in this social, cultural, historical, and political contexts that Joaquín Guzmán Loera, alias "El Chapo" Guzmán, was born in 1957.

The Hippie Movement and the Supply of Drugs from Mexico

At the beginning of the 1960s, the hippie movement in the United States had a profound influence on the market with the least representation in drug production at that time: marijuana. Social movements against the Vietnam War, as well as political strategies within the American government, led to a cultural revolution focused on transforming conservative and rigidly moral society. The consolidation of rock and roll as a cultural expression of a new conception of life, social relations and sexuality, as well as life in peace and in close relationship with nature, influenced the increase in the consumption of all types of drugs by young people, especially marijuana.

This led to a situation in the mid-1960s where figures for marijuana consumption reported by the US government's Substance Abuse and Mental Health Services Administration showed an increase since the beginning of the decade (Valdés, 2013), making it the most used drug among young people in the United States. It has even been estimated that by the end of the 1970s, the United States marijuana market consisted of 30 million smokers who consumed approximately 3660 tonnes of marijuana, of which Mexico exported 2750 tonnes, not including seizures in the border region.

During the 1960s, the main drug producers in Sinaloa began to gain greater prominence and presence in armed confrontations with anti-narcotics police, in drug seizures, as well as in the destruction of drug crops on their properties and in temporary incarcerations (Gómez & Fritz, 2005). In 1969, at the beginning of President Nixon's presidency in the United States, "Operation Interception" was implemented with the aim of closing the borders without prior notice to search all vehicles crossing the border in order to identify and seize drugs.

According to Valdés (2013), in structural terms, the groups involved in drug production during this decade had two main objectives, which consisted of two basic but complex tasks: a) organising the production, harvesting, packaging, and transport of drugs within the national territory and b) ensuring the transport of drugs to the United States.

However, although marijuana production and consumption showed an upward trend during the 1960s and 1970s, opium production and consumption remained stable. In the early 1970s, President Richard Nixon implemented an aggressive policy to control opiate production in Turkey, with France also becoming involved in controlling opium production laboratories linked to imported gum.

These strategies became a strength for both marijuana and opium producers in Mexico, to the point where Mexican heroin covered between 80 and 90 per cent of US demand, increasing by around 70 per cent between 1971 and 1975, resulting in marijuana production alone multiplying several times over. Consequently, this fourth decade of drug production in Mexico led to a significant entrepreneurial and organisational leap in drug trafficking in Mexico (Valdés, 2013).

It should be highlighted that during this decade, binational cooperation operations were implemented against drug production. Operation Condor, in 1975 and 1976, was the most representative in terms of the fight against drugs and environmental impact far from the borders. It was implemented in the Golden Triangle during the presidencies of Luis Echeverría in Mexico and Nixon in the United States, through ground operations, arrests, destruction of drug plantations, and seizures of weapons and drugs (Fig. 5.4).

However, the most notable were the attempts to destroy poppy and marijuana crops from the air using foliar and broad-spectrum herbicides, which were significant in the fight against drugs in Mexico and several Latin American countries over the following decades, and laid the foundations for the emergence of Critical Criminology in Latin America and the emergence of Green Criminology (Goyes & South, 2017).

Fig. 5.4 Destruction of drug plants in the Golden Triangle during Operation Condor in 1975–1976. *Source*: National General Archive (Gobierno de México, 2022)

Shaping the Mexican Pax Narcotica

The strengthening of the market demanding opium derivatives and marijuana led to the expansion of the drug production system in Mexico, which was no longer confined to the "Golden Triangle" of the states of Sinaloa, Durango, Sonora, and Chihuahua. The existence of crops began to be documented in states such as Nuevo León, San Luis Potosí, Michoacán, Guerrero, and Oaxaca, among others (Astorga, 2005; Grillo, 2011; Valdés, 2013). The growing demand for drugs such as marijuana led many agricultural producers to switch from corn to marijuana production. Even in the face of large-scale police operations to stop the production of opium and marijuana, this market proved resilient to efforts to halt production and trafficking to the United States.

Although there is scant information on the consolidation of criminal drug trafficking groups in Mexico during the 1950s and 1980s, it can be argued that the logistics of illegally trafficking drugs into the United States required a broad organisational structure. This involved the production and movement of drugs from states such as Oaxaca and Guerrero in southern Mexico to the border region in the north, the diversification of mechanisms and strategies for crossing the border, such as aeroplanes, tunnels, vehicles with hidden compartments, a system of "mules" with unsuspicious individuals to smuggle smaller shipments across the border, corruption of US customs agents, and the challenges of distributing drugs on American territory.

However, it has been documented that with the start and strengthening of drug production in southern states, it was logistically very difficult to transport the drugs produced to the border region, cross the border, and distribute the drugs in American territory. Therefore, it is considered that agreements between drug producers in southern states and drug producers in the northern region allowed for the consolidation of drug trafficking groups in the northern region of the country, as they purchased drug production from southern states while supplying American demand for drugs with production from both areas of the country. This meant that producers in the southern states planted and harvested the drugs to sell them to producers in the north, who had greater operational capacity to cross the border and distribute the drugs in the United States.

Although it is unknown how agreements between leaders were made, the 1970s were characterised by laying the foundations for the emergence of drug cartels as we know them today. According to historical records, the size of the demand for drugs in Mexico required a well-organised agro-industrial production and logistics business, which was in turn strengthened by impunity, government protectionism at the local level, and collaborative networks at the national and international levels, mainly in the United States. In 1975, the arrest of Alberto Sicilia Falcón represented a change in American efforts to end drugs during the Nixon administration. This arrest marked a change in anti-drug strategies in Mexican territory, avoiding the arrest of small producers and dealers and arresting the leaders (Grillo, 2011).

The DEA (Drug Enforcement Administration) informed the Mexican police about Alberto Sicilia Falcón's activities, and he was arrested in Mexico City. Bank accounts in Switzerland worth $260 million were found, along with evidence linking him directly to politicians and celebrities. Furthermore, in his statement following his arrest, he claimed that he was working on a CIA (Central Intelligence Agency) operation to use drug money to buy weapons for Central American rebels. He even claimed to have been trained by the CIA at Fort Jackson, Florida (Grillo, 2011).

This arrest marked the first time a foreign drug capo had been arrested in Mexico. According to Valdés (2013), the Falcón case served to demonstrate that Cuban and American gangsters with experience in criminal activities, cross-border trafficking networks, and money laundering fuelled the expansion of drug trafficking in Mexico. This was evident in the way Mexicans learned about this global business, which was decisive in the ways Mexican drug traffickers operated in the following decades.

By the end of the 1970s, the Sinaloa organisation had expanded its control along the entire Pacific coast, from the states of Oaxaca and Guerrero (as marijuana producers), through Jalisco and Nayarit, to Sinaloa, Sonora, and Baja California, producing opium in the Golden Triangle (see Figure) with leaders such as Pedro Avilés Pérez, also known as "El León de la Sierra" (The Lion of the Sierra), Ernesto Fonseca Carrillo, "Don Neto", and Eduardo Fernández Juárez, "Don Lalo".

Meanwhile, in the border region of Ojinaga in the state of Chihuahua, the area known as "Plaza de Ojinaga" operated by a drug trafficker named Pablo Acosta, after several local conflicts over control of that border crossing, it became established as an ideal geographical location for drug trafficking to the United States, particularly to Presidio, Texas, because it was a small town and remote. In addition to producing marijuana, cocaine from Colombia arrived there with the approval or endorsement of the Sinaloa organisation.

This approval by the Sinaloa organisation to traffic drugs produced in Colombia through Ojinaga was due to the recommendation of Ernesto Fonseca Carrillo, known as "Don Neto", or Miguel Ángel Félix Gallardo, that Pablo Acosta's Ojinaga Plaza[1] was safe for bringing cocaine into the United States. However, despite the fact that there was no direct conflict between the Sinaloa organisation and the Ojinaga Plaza, Sinaloa's endorsement was conditional on a kind of "supervising manager" known as Amado Carrillo Fuentes, who was the link between the two organisations and who, after Pablo Acosta's death, moved his base of operations to Ciudad Juárez and became known as the "El Señor de los Cielos" (The Lord of the Skies) (Valdés, 2013).

Once the large drug production and trafficking groups in Mexico had consolidated their power in the 1970s, this period was characterised by the centralisation of drug production, the formation of "plazas" (drug trafficking territories), the

[1] The term "plazas" refers to the temporary or permanent control of drug trafficking in a region, city, or territory, generally by an individual and his or her criminal group.

cessation of conflicts between drug traffickers and Mexican state forces due to corruption and impunity, both at the police and military levels, and the consolidation of agreements with Mexican and American customs agents for the crossing of drugs across the border. Until 1980, the territorial expansion of drug production alternated with the emergence of natural areas dedicated to environmental conservation by presidential decree. According to records, at the end of that decade, there were 93 areas designated for environmental conservation in Mexico (Fig. 5.5) (see Supplementary Material 5.1).

Historical records do not allow for the exact distribution of all drug cultivation sites to be traced. However, an analysis of the distribution by state reveals that, since the decree establishing the first environmental conservation areas, there have been territories that were already occupied by drug traffickers.

Environmental Harms of Drug Production in This Period

Although there are no records or studies on the environmental harms caused by drug production during this period in Mexico, the main drug production areas were located far from urban centres, mainly in rural and wild environments. One of the

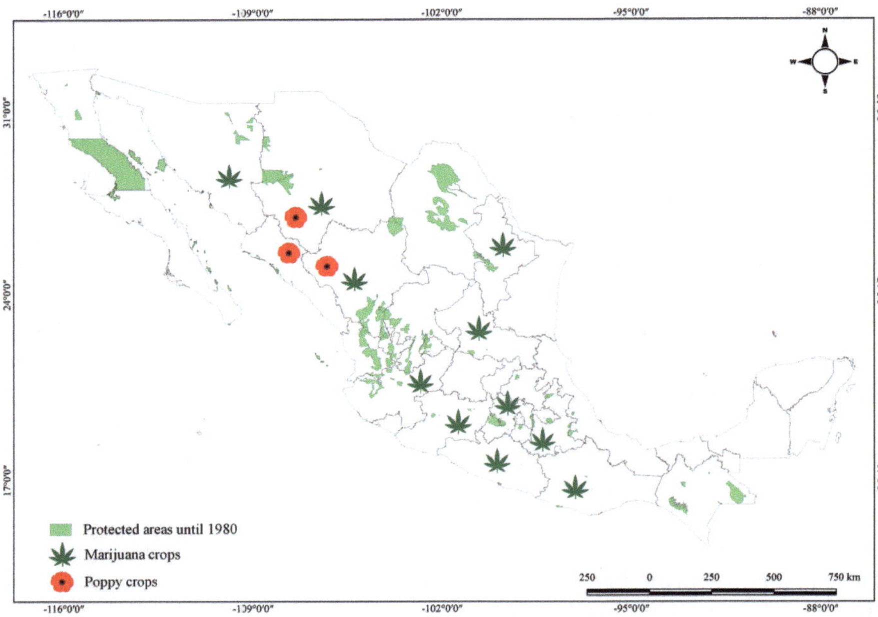

Fig. 5.5 Geographical distribution by state of drug production and the first areas for environmental conservation in Mexico (until 1980). *Note*: (**a**) In this historical period, there were no institutionalised (Natural) Protected Areas, there were only presidential decrees for the conservation of some natural areas, (**b**) the images of marijuana and poppy crops do not show the exact location, they only show the main drug produced in that state

main characteristics of the "Golden Triangle" was its location in the mountainous areas of the Sierra Madre Occidental or in forested areas of Oaxaca and Guerrero, which limited government action but also made the environmental harm caused by this activity invisible.

Strategically, the cultivation of marijuana and poppies prevented the deforestation of large areas of forest, as it was carried out in many small spaces free of forest vegetation to camouflage the marijuana and poppies in case of overflights by security agencies to identify drug plantations.

However, the potential local impact near drug plantations cannot be dismissed, as documented in other global studies (United Nations Office on Drugs and Crime [UNODC], 2022). For example, drug plantation surveillance camps could have some environmental impact at the ecosystem level, as firewood was mainly used for cooking and heating, which involved the use of dry and fresh wood (illegal logging), the generation of waste and water pollutants such as cooking oil, the use of latrines or open defaecation, as well as bathing in waterways that could contaminate them with soap residues. Another possible alteration to local ecosystems was the use of fertilisers to ensure healthy marijuana and poppy plants. In addition, due to the fertility of the soil in these regions, this could possibly lead to a loss of soil fertility.

In terms of the impact on biodiversity, there was illegal hunting of wildlife, mainly deer, wild boar, rattlesnakes, and other animals for human consumption. Meanwhile, for safety reasons, crop guards hunted animals such as bears, coyotes, pumas, and jaguars out of fear or caution against attacks by wildlife.

From the perspective of green criminology, environmental harms can be classified into three types according to its origin (Carpio-Domínguez, 2023): (a) harms caused by civil society, i.e., individuals who cause environmental harm for criminal activities and/or subsistence purposes; (b) harms caused by companies and industries for the accumulation of capital, and (c) harms caused by government institutions through productive practices, government responses, or the modification of laws and approval of permits to the industrial sector and bad legislative and public policy practices that favour or contribute to environmental harms.

Drug production as it was carried out until the 1970s can be considered environmental harm for subsistence. However, it should be noted that government action in the fight against drugs during the 1960s and 1970s consisted of crop fumigation from light aircraft belonging to the Attorney General's Office. These fumigations were carried out with broad-spectrum foliar herbicides, so the environmental impact of the actions taken to combat drugs cannot be dismissed; furthermore, due to the lack of studies, the environmental impact caused by these actions is unknown. In this regard, government action against drugs through the use of broad-spectrum herbicides on drug crops and aerial fumigation strategies could be considered environmental harm in areas dedicated to environmental conservation.

Conclusions

The emergence of drug trafficking in Mexico is a multifactorial phenomenon that must be understood within and from its socio-historical, cultural, and economic contexts. Drug production at the national level went from being a leisure activity and for personal consumption to representing a transnational economic and agro-industrial sector. This brief historical overview documents the phases that occurred during the emergence of drug trafficking, from drug production by people of Chinese origin, to the emancipation of Mexican drug production and the consolidation of drug trafficking networks that were decisive in the following decades.

During this period, drug production and the consolidation of drug trafficking groups alternated and shared a geographical, social, historical, cultural, and environmental space with the emergence of areas designated for environmental conservation. This is evident in how drug production requires suitable biophysical spaces for the planting and harvesting of poppy and marijuana, particularly considering that Mexico's biogeographical characteristics facilitated and influenced the conditions in which drug production developed.

The biogeographical characteristics of the Sierra Madre Occidental in northern Mexico, mainly in the states of Sinaloa, Chihuahua, Durango, and Sonora, facilitated drug production due to the accessibility of water, soil fertility, and clearly defined seasons; while socially, it highlighted the availability of labour for the entire drug production process (planting, cultivation, and harvesting), as well as the economic necessity due to the remote location of the areas where drugs were produced, coupled with the government's neglect of the basic needs of these rural and indigenous populations. All these factors contributed to the entrenchment of drug trafficking.

Despite internal conflicts among producers, the fledgling drug trafficking business managed to consolidate itself and meet the growing demand for drugs. It is also important to highlight the significant influence of the United States on the criminal dynamics of drug production in Mexico, influenced by sociocultural phenomena such as the hippie movement and the increase in demand for marijuana and the sustained consumption of opium and its derivatives for decades, which required the massive influx of drugs into US territory; while in the governmental sphere and its institutions in both countries, the prohibition, extermination, and criminalisation of drug production and consumption were evident.

This period analysed shows the prelude to the consolidation of the large drug cartels in Mexico from the 1980s onwards and allows to understand that drug trafficking in its early days was characterised by the emancipation of drug production from Chinese families in Mexico. Resistance to prohibition, the creation of pacts and alliances between producers, the emergence of leaders to coordinate actions, violence as a means of dominating territories and ensuring loyalty, the relationship between drug trafficking leaders and Mexican and US government authorities, and the profitability of a business with economic gains for all members of the market.

Finally, drug production predates the emergence of (natural) protected areas, which meant that in the following decades, the decrees establishing protected areas coincided with territories historically occupied by drug producers and under the control of drug cartels.

References

Arroyo-Quiroz, I., & Wyatt, T. (2019). Wildlife trafficking between the European Union and Mexico. *International Journal for Crime, Justice and Social Democracy, 8*(3), 23–37. https://doi.org/10.5204/ijcjsd.v8i3.1243

Astorga, L. (2005). *El siglo de las drogas. El narcotráfico, del Porfiriato al nuevo milenio*. Plaza y Janés.

Bonnie, R., & Whitebread, C. (1999). *The marihuana conviction: A history of marihuana prohibition in the United States*. The Lindesmith Center. https://cAnnAlib.eu/wp-content/uploads/2022/05/the-Marijuana-Conviction-A-history-of-marijuana-prohibition-in-the-United-States-1999.pdf

Campos, I. (2010). Degeneration and the origins of Mexico's war on drugs. *Mexican Studies/Estudios Mexicanos, 26*(2), 379–408. https://doi.org/10.1525/msem.2010.26.2.379

Capó-Valdivia, Z. (2015). Clandestinidad, fiscalidad y tolerancia: el opio en el discurso oficial, Mexicali, México 1915-1916. *Culturales, 3*(2), 77–111. https://www.scielo.org.mx/scielo.php?script=sci_arttext&pid=S1870-11912015000200003

Carpio-Domínguez, J. (2021). Crimen Organizado (narcotráfico) y conservación ambiental: el tema pendiente de la seguridad pública en México. *Revista CS, 33*, 237–274. https://doi.org/10.18046/recs.i33.4076

Carpio-Domínguez, J. (2023). *Criminología verde para México: Desarrollo de una perspectiva verde en la herencia criminológica mexicana*. Universidad Autónoma de Tamaulipas-Tirant Lo Blanch. https://doi.org/10.29059/LUAT.317

Carpio-Domínguez, J., Arroyo-Quiroz, I., Villarreal-Martínez, M., & Castro-Salazar, I. (2022). Tigers of Mexican drug traffickers, national statistical review and a case study through network analysis of the possession of a tiger (*Panthera tigris*) in northeastern Mexico. *Forensic Science International: Animals and Environments, 2*(2022), 100039. https://doi.org/10.1016/j.fsiae.2021.100039

Carpio-Domínguez, J., Vargas-Orozco, C., & Villarreal-Sotelo, K. (2023). Redes de tráfico ilegal de vida silvestre en el noreste mexicano, un análisis desde la criminología verde. *Revista Española De Investigación Criminológica, 21*(2), e789. https://doi.org/10.46381/reic.v21i2.789

Carpio-Domínguez, J., Vargas-Orozco, C., Villarreal-Sotelo, K., Santillana-Cantú, R., & Hernández-Rodríguez, I. (2018). Percepción Criminológica de la Posesión de Animales por el Narcotráfico en Tamaulipas, el Zoológico del Crimen Organizado. *Letras Jurídicas, 26*, 1–18. https://dialnet.unirioja.es/servlet/articulo?codigo=6954735

Correa-Cabrera, G. (2016). Militarización y violencia en Tamaulipas. Taller sobre Seguridad Ciudadana en México. In *Atlas de la Seguridad y la Defensa de México* (pp. 181–189). CASEDE. https://www.casede.org/PublicacionesCasede/Atlas2016/Guadalupe_Correa.pdf

De la Maza, R. (1999). Una historia de las Áreas Naturales Protegidas en México. *Gaceta Ecológica, 51*, 15–45. https://paot.org.mx/centro/ine-semarnat/gacetas/GE51.pdf

Enciso, F. (2010). Los fracasos del chantaje. Régimen de prohibición de drogas y narcotráfico. In A. Alvarado & M. Serrano (Eds.), *Los grandes problemas de México. Seguridad nacional y seguridad interior* (pp. 61–104). El Colegio de México. https://doi.org/10.2307/j.ctt1657t6f.7

Gobierno de México. (2022). *Los años germinales de la política de guerra contra las drogas en México*. Gobierno de México. https://www.gob.mx/agn/articulos/los-anos-germinales-de-la-politica-de-guerra-contra-las-drogas-en-mexico?idiom=es

Gómez, M., & Fritz, D. (2005). Con la muerte en el bolsillo. In *Seis desaforadas historias del narcotráfico en México*. Planeta.
González-Félix, M. (2002). Empresarios y gobierno en el Distrito Norte, 1902-1920. In Velázquez-Morales, C. (Coord.) *Baja California: Un presente con historia (Tomo II)* (pp. 13–28). Universidad Autónoma de Baja California.
Goyes, D., & South, N. (2017). Green criminology before 'green criminology': Amnesia and absences. *Critical Criminology, 25*, 165–181. https://doi.org/10.1007/s10612-017-9357-8
Grillo, I. (2011). *El Narco: Inside Mexico's criminal insurgence*. Bloomsbury Press.
Mack, A., & Joy, J. (2000). *Marijuana as medicine? The science beyond the controversy*. National Academies Press. https://www.ncbi.nlm.nih.gov/books/NBK224394/
Musto, D. (1997). Opium, cocaine and marijuana in American history. In K. Larry Gaines & P. Kraska (Eds.), *Drugs, crime, and justice: Contemporary perspectives* (pp. 21–33). Waveland Press Inc.
Paoli-Bolio, I. (2008). Evolución del narcotráfico en México. *Bien Común, 163*, 98–110. https://biblat.unam.mx/es/revista/bien-comun/articulo/evolucion-del-narcotrafico-en-mexico
Pérez Montfort, R. (1997). Fragmentos de historia de las "drogas" en México, 1870-1920. In Pérez Montfort, R. (Coord.) *Hábitos, normas y escándalo: Prensa, criminalidad y drogas durante el porfiriato tardío* (pp. 145–204). Centro de Investigaciones y Estudios Superiores en Antropología Social/Plaza y Valdés.
Rodríguez, O. (2022). ¿De dónde salieron y a dónde se fueron? Migración interna de regiones de alta violencia en México en las últimas dos décadas. *Revista EURE - Revista De Estudios Urbano Regionales, 48*(144), 12. https://doi.org/10.7764/EURE.48.144.12
Sandos, J. (1984). Northern separatism during the Mexican revolution: An inquiry into the role of drug trafficking, 1910-1920. *The Americas, 41*(2), 191–214. https://doi.org/10.2307/1007456
United Nations. (1912). International opium convention. In *League of nations, treaty series* (Vol. 8, p. 187). United Nations. https://treaties.un.org/pages/ViewDetailsIV.aspx?src=TREATY&mtdsg_no=VI-2&chapter=6&Temp=mtdsg4&clang=_en
United Nations Office on Drugs and Crime (UNODC). (2022). *World drug report. Booklet 5-drugs and the environment*. United Nations. https://www.unodc.org/res/wdr2022/MS/WDR22_Booklet_5.pdf
Valdés, G. (2013). *Historia del narcotráfico en México*. Editorial Aguilar.

Chapter 6
Pax Narcotica: Government, Organised Crime, and Environmental Governance

Introduction

To understand the consolidation of drug trafficking and the major drug cartels in Mexico and their relationship with areas designated for environmental protection, it is necessary to clarify that policies for the protection of natural areas were implemented randomly and in response to the specific needs of each region. It was not until 2000 that protected areas were formalised and institutionalised through the creation of the Comisión Nacional de Áreas Naturales Protegidas (CONANP) (National Commission for Protected Natural Areas).

By the 1980s, groups dedicated to drug production already had a functional structure in place to supply a demanding market and were under the protection of government forces in Mexico. Since the 1970s, the enormous amounts of money generated by drug trafficking and the power wielded by leaders in the territories were characteristics that governments at all levels (federal, state, and municipal) could not ignore. This led to sophisticated political-economic relationships centred on the acquisition of money and power, although a direct relationship or agreements between leaders, such as the Sinaloa Organisation and the political sphere, were never formally recognised. However, the evidence would confirm these relationships.

According to Valdés (2013), during this time questions arose about how it was possible that government entities at all levels were unable to combat drug trafficking. Questions arose about whether, at the federal level, the president of the republic was part of the emerging drug trafficking structure in Mexico, or if the president was unaware of or not part of this structure, then the secretaries of security, governors, municipal presidents, ministers of justice, and police commanders set the agenda for drug production and transport through legal protection and impunity.

Supplementary Information The online version contains supplementary material available at https://doi.org/10.1007/978-3-032-18218-0_6.

J. L. Carpio-Domínguez et al., *Drug Trafficking and Protected Areas in Mexico*,
https://doi.org/10.1007/978-3-032-18218-0_6

This chapter presents a documentary analysis based on the works of Astorga (2005) and Valdés (2013) to provide a diachronic guide to the chapter due to its importance in the historiography of drug trafficking in Mexico. This chapter does not aim to offer an in-depth historical analysis of the emergence and consolidation of drug trafficking, but it does recover historical data that allows to contextualise and connect the development and consolidation of drug trafficking with the emergence of areas for environmental conservation, as well as to rescue records on the environmental impacts caused by the presence of these criminal groups in these areas and the government's fight against drug trafficking.

Through documentary analysis, this chapter examines how the fragmentation of Mexico's main cartel during the 1980s and 1990s created a new scenario for drug trafficking in Mexico in the following decades. This chapter analyses the importance of national and foreign government efforts in the fight against drug trafficking, as well as the environmental impact caused by drug trafficking and the efforts of governments to eliminate drug production and the associations between drug traffickers and high-level public servants in Mexico.

Models of Drug Trafficking–Government Relations and the "Mexican Variant" in the Pax Narcotica

In the literature on organised crime, Bailey and Godson (2000) proposed two models for understanding the relationship between organised crime and the political sphere (Fig. 6.1): a) the "Centralised Systemic" model, which considers a system with a central guide or leadership that facilitates and establishes the link between the Mexican political system and organised crime; this is a "top-down" model in

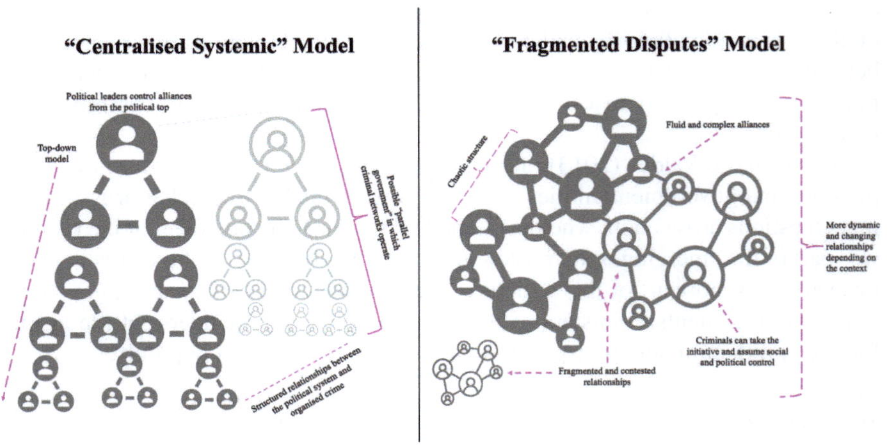

Fig. 6.1 Models between the political system and organised crime. *Source*: the authors with data from Bailey and Godson (2000)

which political leaders control networks of alliances from the top down; and even operating in a shadow system as a parallel system of government, leading criminal activities without being formally linked to the government; and b) the "Fragmented Disputes" model frames a more fluid and complex set of relationships between the political system and organised crime, implying more dynamic and changing relationships and alliances depending on the variety of circumstances in each context.

The period from 1980 to 2006 is known as the "pax narcotica" or "pax narca" period and refers to a period of alliances between drug traffickers and the government sector, in which government action to enforce the law on drug trafficking practices was selective and geared towards particular interests. Thus, there was government protectionism towards drug trafficking leaders, and therefore arrests of drug traffickers, destruction of drug crops, and seizures of weapons and drugs were only implemented when it was necessary to comply with a political agenda, break agreements between leaders, or due to political pressure in the bilateral agreement between the United States and Mexico.

To understand this period of pax narcotica, it is necessary to highlight that drug trafficking received benefits from the government's institutional apparatus, Valdés (2013) refers to as tangible and intangible assets. About tangible assets, the main one is money, which was evident in the form of operating permits and relative protection in the form of fees, which were paid directly to Mexican security agencies, leading to the enrichment of public servants in public security institutions.

While intangible assets refer to empowerment over territories or plazas under an agreement with government institutions, among the intangible assets was the containment of violence. According to Valdés (2013), the absence of violence was due to: (a) the relatively peaceful subjugation of border territories by the Sinaloa Organisation, and (b) the Mexican government's own strategy of "managing" organised crime by plazas.

Another consequence of the consolidation of drug production in Mexico and the alliances between drug traffickers and public servants was the monopolisation of the drug market by the Sinaloa Organisation, with the support of public security institutions. According to historical records, drug trafficking infiltrated all offices of the Attorney General's Office through economic incentives, informal agreements, protectionism, and impunity. Thus, being a delegate or prosecutor ensured millions of dollars in benefits, positions were sold for thousands of dollars, and according to Valdés (2013), representing the Attorney General's Office was a farce; in fact, drug traffickers were being protected.

To understand the dynamics of the relationship between drug traffickers and public servants in public security, the following excerpt from Valdes (2013, p. 170) is worth noting:

> Let's do a big illegal business together. I, the government, will control the board of directors and set the rules of the game because I am the State; you, drug trafficking corporations, will operate with my blessing and protection, in exchange for sharing a slice of the profits with me. In addition, you must promise to behave like a 'decent' organised crime group, that is, you will behave yourselves and not act against institutions or society. If any of you cause too much trouble, do not respect my rules, or I find myself uncomfortable due to external

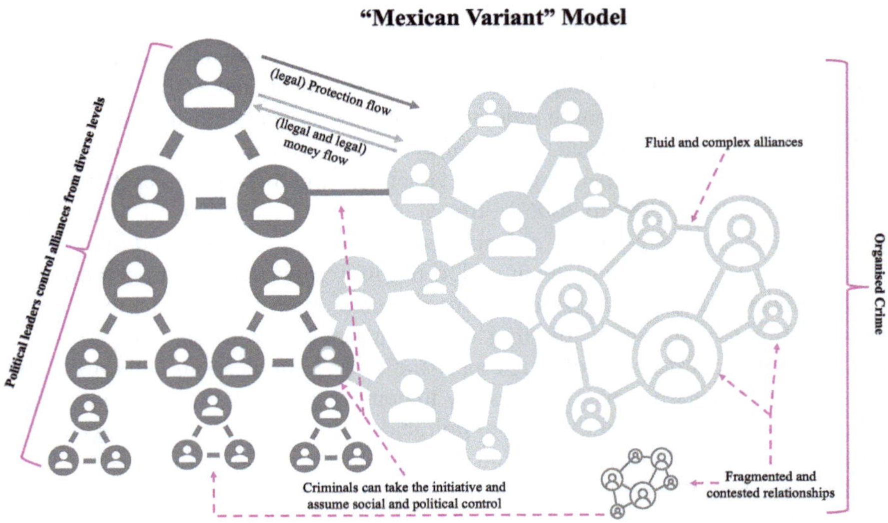

Fig. 6.2 Mexican Variant model of organised crime. *Source*: the authors with data from Valdés (2013)

> pressures, I will have to act against some of you, but do not worry: the damage will be limited.

This excerpt clearly defines what pax narcotica or pax narca is. The benefits received by both parties (drug traffickers and the government) consolidated the monopolisation of drug trafficking and the ways in which it was carried out over the following decades. Although the weakening of the government's ability to enforce the rule of law is generally associated with the presence of criminal groups engaged in drug trafficking, as evidenced by other criminal groups in other parts of the world, organised crime theorists referred to the drug trafficking–government relationship as the "Mexican Variant" model (Valdés, 2013, p. 173).

In summary, the Mexican Variant model consists of criminal organisations that operate without the inconvenience of having the authorities against them, through a pact with the State, which only pursues them sporadically and selectively and allows them to operate freely without hiding, travel without problems on the country's roads, integrate vertically and horizontally into society and government institutions, and put police and other public security agents at their service (Fig. 6.2).

Finally, the consolidation of drug trafficking in Mexico was due to two main causes: a) growing demand for drugs in the United States; and b) the Mexican state facilitated, protected, and supported the consolidation of these criminal groups against all logic of the rule of law, and in the process facilitated their spread throughout the national territory, including areas designated for environmental conservation, which were (and remain) strategic territories for drug production, processing, and trafficking.

Overview of Drug Trafficking in the 80s

In the early 1980s, drug trafficking in Mexico was dominated by the Sinaloa Organisation, which, at that time, was known by public security agencies and the media as the Guadalajara Cartel because its leaders had moved to the capital of Jalisco after Operation Condor (1976–1977). According to historical records, the leaders centred their operations in Guadalajara due to their close relationship with the then director of the Federal Security Agency, who provided them with shelter and protection through its extensive network of political and police contacts in the state of Jalisco.

According to Aguilar (2009) and Valdés (2013), after Operation Condor, the Federal Security Agency recommended to Félix Gallardo and Ernesto Fonseca, alias "Don Neto" (leaders of the Sinaloa Organisation): (1) to put an end to internal conflicts within the Organisation, (2) set up a base of operations in the United States, (3) to leave Sinaloa, and (4) to find refuge in Guadalajara. These recommendations led to the consolidation of the Guadalajara Cartel through government support. On the one hand, the emerging cartel provided the manpower and blood (traffickers and sicarios), while the government provided intelligence, coordination, and protection.

The Guadalajara Cartel, led by Miguel Ángel Félix Gallardo (a former police officer) and Ernesto Fonseca, alias "Don Neto", was the most important cartel during this period because it controlled the largest areas of marijuana and poppy production. It operated in the states of Oaxaca, Guerrero, Michoacán, Jalisco, Nayarit, Sinaloa, Sonora, and the two drug-producing strongholds: Durango and Chihuahua. They also controlled the main border crossings in Baja California (Tijuana, Tecate, and Mexicali) and Sonora (San Luis Río Colorado and Nogales), and they benefited financially from their association with Pablo Acosta as head of the Ojinaga plaza in Chihuahua.

Therefore, the Guadalajara Cartel was the dominant drug trafficking group at the national level, composed of family associations such as the Caro Quintero, Carrillo Fuentes, Guzmán Loera, and Arellano Félix families, among others with less direct family ties but with strongholds such as the Salcido Uzeta family in Sinaloa, the Beltrán Leyva family in Sonora, the Herrera family in Durango, the Lupercio Serratos family in Guadalajara, the Alba Álvarez family in Guerrero, and the Díaz Parada family in Oaxaca, all under the leadership of Miguel Ángel Félix Gallardo, alias "Jefe de Jefes" (Boss of Bosses), and Ernesto Fonseca Carrillo, alias "Don Neto" (see Supplementary Material 6.1) (Salazar-Pérez & Fernández Velázquez, 2016).

To understand the relevance of this criminal group, it is necessary to highlight that, in addition to government protectionism, drug trafficking as an organisation performed public security functions for other crimes. Citizens turned to the cartel leaders and their subordinates to seek justice, accelerate criminal investigations, and reduce rates of violence, theft, and assault. Paradoxically, this was achieved in less time than government action. This meant two things: (1) the cartel, in association with the government, administered justice, and (2) the citizens had complete trust in the cartel to address social and public security issues.

In this context of the pax narcotica, the members of this drug cartel were characterised as social benefactors, supporting the development of the localities where drugs were produced and where the leaders were based; and with government support, they guaranteed public security. These activities became a characteristic of this criminal group, so that its public image did not always match the illegal activities it carried out (drug production, homicides, injuries, etc.) and it had a profound impact on social protection and the expansion of its territories, including areas for environmental conservation and nearby territories where drugs were produced, generating jobs.

Among the most notable characteristics of this period was the technological strengthening of drug production (poppy and marijuana) at the industrial level, which went from being seasonal (dependent on the seasons of the year) to one with irrigation systems, the rental and purchase of land for production, as well as an increase in labour. This was evident in 1984 with the raid on the "El Búfalo" ranch in the municipality of Allende in Chihuahua, owned by Rafael Caro Quintero of the Guadalajara Cartel, as part of "Operation Pacific" led by the United States and Mexican governments (Santana, 2018).

This 12-square-kilometre marijuana ranch, nestled in the Chihuahuan Desert, demonstrated the ability to produce drugs even in desert conditions through irrigation systems. Around 12,000 labourers worked on the ranch, mainly from the state of Sinaloa, but also from Sonora, Guerrero, Oaxaca, Chiapas and even Guatemala. They were guarded by armed men and fed basic foods (beans, tortillas, potatoes, and coffee) (Fig. 6.3).

During the operation on this ranch, 11,000 tonnes of marijuana were destroyed, considering that at that time a pound of marijuana cost 1000 (US dollars). In addition to the lack of trust in the police to report drug production activities on this ranch, there were previous reports of marijuana and poppy plantations in other areas within the state of Chihuahua, to which the governor responded that "they were just people's hallucinations" (Astorga, 2005, p. 133).

Mexico as a Bridge for Colombian Cocaine Trafficking

In this context, during the 1980s, there was a significant change in the way drug trafficking evolved in Mexico. The criminal associations of the Guadalajara Cartel not only distributed drugs nationally and in the United States, but also in other Central American countries.

Colombia began to make its first appearances in the history of drug trafficking in Mexico through agreements between the leaders of the Guadalajara Cartel and drug producers in Colombia. This agreement had already been preceded by the entry of cocaine through the Ojinaga Plaza with Pablo Acosta in 1970 through an agreement between the Medellín Cartel, the Cali Cartel, and the Guadalajara Cartel, which

Fig. 6.3 Operation Pacific at the "El Búfalo" ranch in Allende, Chihuahua, owned by Rafael Caro Quintero of the Guadalajara Cartel in 1984. *Source*: El Heraldo de Chihuahua (2021). *Notes*: (**a**) day workers inspected during the operation; (**b**) huts in which the day workers slept; (**c**) marijuana plants piled up for burning; (**d**) burning of marijuana, the height of all the marijuana piled up is estimated to be approximately five metres

charged up to half of the profits for crossing cocaine through Mexican territory. During this decade, cocaine trafficking became consolidated and officially began to form part of the main group of drugs trafficked from Mexico to the United States.

Although cocaine originated in Bolivia and Peru, Colombians knew how to turn it into an international and then global business. Colombians imported cocaine paste from Peru and Bolivia, chemical elements for processing from Europe, and weapons to support the business from the United States via Panama. The cocaine was then refined locally (Melo, 1998). It should be noted that cocaine was initially refined in city centres in improvised home laboratories and later moved to rural and jungle areas, which led to the incursion of illegal activities related to drug production in wild environments in Colombia (Morales, 2017).

The emergence of cocaine on the drug scene had a considerable impact on the American population and on the ways in which drug use was perceived at the social and governmental levels. Historically, the use of marijuana and opium derivatives was associated with African American, Asian, and Latino populations. However, cocaine was characterised as a drug consumed mainly by the white population. In the early 1980s, it was estimated that 80% of cocaine users were white people (Baum, 1996).

Unlike other drugs such as marijuana and opium derivatives, cocaine stimulated the mind and increased physical energy, leading to greater performance at parties and work. It was considered a clean, glamorous, sexy, and fashionable drug and became an icon of parties and private gatherings, as well as a symbol of social and economic status (Grillo, 2011). Cocaine trafficking networks to the United States facilitated the entry of more than 50 tonnes by sea through the Gulf of Mexico.

The consolidation of cocaine in the drug trade had a considerable impact on states such as Florida, with Miami becoming the southern hub of the United States for cocaine importation and consumption, leading to an increase in money laundering and linked to the urban development of the city. However, this consolidation process was not without violence. The increase in drug-related homicides and armed clashes between Cuban and Colombian drug traffickers caused the deaths of innocent victims.

To tackle this new drug crisis in the United States, President Ronald Reagan (1981–1989) gave the order to resume the war on drugs, which led to US Army and Navy units locating, tracking, and pursuing suspected drug traffickers, and notifying the Coast Guard for law enforcement. This anti-drug operation (known as the South Florida Task Force) increased seizures of ships carrying cocaine from Colombia, strengthening alliances between the Guadalajara Cartel and Colombian cartels and redirecting trafficking routes to the continental shelf to transport drugs by land across Mexico's northern border (Rosas, 2025).

The Mexican experience in producing and trafficking drugs to the United States since the 1930s meant that Colombian drug traffickers needed alliances. Juan Manuel Matta Ballesteros, a Honduran with a history of illegal activities related to drug trafficking in the United States, was the link between the Guadalajara Cartel and the Colombian cartels. This agreement consisted of using Mexican drug trafficking networks to the United States to bring in cocaine produced in Colombia, generating profits of approximately five million dollars a week for the Guadalajara Cartel and Juan Manuel Matta Ballesteros.

According to Valdés (2013), this agreement represented the third wave of drug trafficking in Mexico. The first was the demand for and supply of opium to the United States during the Second World War in the 1940s and 1950s, the second was the demand and supply of marijuana during the 1960s with the hippie movement and protests against American involvement in the Vietnam War, and finally, the emergence of cocaine during the 1980s as a new drug trafficked to the United States.

The Fading of the Golden Era of Mexican Drug Trafficking, the Loss of the Leaders

During this time, the Guadalajara Cartel was the dominant criminal group in Mexico and was considered an international criminal group strengthened by government protectionism at the micro and macro institutional levels. However, the operation carried out at Rancho "El Búfalo" in 1984 had left Rafael Caro Quintero deeply angered. Confident in their government protection, Caro Quintero and his leaders Ernesto Fonseca Carrillo and Miguel Ángel Félix Gallardo decided to punish the main leader of the operation, DEA agent Enrique Camarena Salazar.

On 7 February 1985, outside the United States Consulate in the city of Guadalajara, an armed commando unit from the Guadalajara Cartel, which also included judicial police officers from the state of Jalisco, kidnapped Enrique Camarena Salazar, alias "Kiki Camarena" and pilot Alfredo Zavala (a pilot for the Secretariat of Agriculture and Hydraulic Resources who had informed Camarena of the existence of the drug-producing ranch). They were tortured and murdered.

The bodies of Kiki Camarena and Alfredo Zavala were found on 7 March on a ranch in the state of Michoacán (owned by a former local deputy), causing outrage in the United States and marking the beginning of the decline of the golden age of Mexican drug trafficking. Five days after the bodies of Kiki Camarena and Alfredo Zavala were found, Caro Quintero fled to Costa Rica. The Mexican government and the DEA began their pursuit, and he was arrested on 4 April 1985. After his arrest, he was identified as the third in command within the Guadalajara Cartel, the second in command was Miguel Ángel Félix Gallardo, and the boss of both was Ernesto Fonseca Carrillo.

Astorga (2005) recovers an excerpt from an interview in which Caro Quintero requested to be judged as a drug trafficker rather than a murderer and argued that he had never had any problems with the law, emphasising that he had always worked in the mountains "where there was no law" (p. 138). This excerpt from the interview highlights the importance of natural and wild spaces for drug production, as their remoteness from urban centres and government action meant they were considered lawless territories, or at least places where justice did not reach, and ideal for drug production (Carpio-Domínguez, 2021).

Another strike against Mexican drug trafficking was the arrest of Ernesto Fonseca Carrillo, "Don Neto" on 7 April 1985 in the Mexican port of Puerto Vallarta in the state of Jalisco. The arrest was carried out peacefully at a residence owned by the municipal police chief. With two of the three leaders of the Guadalajara Cartel under arrest, it became clear that the police protection they had bought over the years was not enough to hide them from US persecution, which was represented and carried out by the Mexican authorities themselves (Valdés, 2013). Following Fonseca Carrillo's arrest, organisational adjustments were made within the Attorney General's Office, dismissing the commander of the Eagle Group and 13 agents from the Federal Judicial Police and the Jalisco Judicial Police.

In April 1989, Miguel Ángel Félix Gallardo was peacefully arrested by Mexican and US authorities in the city of Guadalajara, Jalisco. The arrest was carried out by an acquaintance of Félix Gallardo, who was deputy director of the Federal Judicial Police through the Attorney General's Office. According to testimonies and historical records of the interrogations, the arrest between Félix Gallardo and the public servants was due to political pressure from the binational agreement between Mexico and the United States and that in reality there was no evidence to link him to the crime, so with the start of the new presidential term in Mexico, he would be released.

However, Félix Gallardo's arrest was followed by other arrests throughout the country related to this case, mainly of public servants who had supported and protected Félix Gallardo's drug trafficking operations for many years. In Sinaloa, the deputy head of the anti-drug trafficking campaign and the head of the State Judicial Police, the head of the Culiacán Judicial Police, a Federal Highway Police Officer, the Regional Commander of Tamaulipas, and an officer based in Nuevo León were arrested.

With the start of Carlos Salinas de Gortari's 6-year term (1988–1994), the agreements and promises made to Felix Gallardo were not fulfilled. The new president's approach to collaboration with the US government, known as "the spirit of Houston" (after a meeting with President George Bush), led to the continuation of the US line of work in relation to drug control. In this scenario, in which it seemed that the Mexican government had regained control over drug trafficking, Valdés (2013) defines it as follows: "a bossy government, decapitated organised crime and a fascinated DEA" (p. 210). However, no one at the time imagined what would happen next in the drug trafficking scene in Mexico: fragmentation and violence.

The capture of the three leaders led to the break-up of the Guadalajara Cartel into factions, although historical records mention that Félix Gallardo left the structure and working methods of the Guadalajara Cartel intact (Table 6.1). The organisation of the Guadalajara Cartel's work by territories was supposed to facilitate the continuation of its activities once the leaders were no longer at the forefront of the cartel.

These territories were later consolidated into family organisations that dominated the drug trafficking scene in the following decades, while others disappeared, annexing themselves to more dominant ones in a scheme similar to the feudal system. This division by territory, which was reported by the media in the late 1980s, did not survive for long (see Table 6.1). In the following years, adjustments within the Guadalajara Cartel led to the prevalence of family-headed organisations (Fig. 6.4):

Table 6.1 (Alleged) distribution of drug trafficking plazas by Miguel Ángel Félix Gallardo in the late 1980s

Municipalities (drug trafficking plaza)	State	Leadership
Tecate	Baja California	Joaquín Guzmán Loera
Tijuana		Jesús "Don Chuy" Labra
Mexicali		Rafael Chao
Ciudad Juárez	Chihuahua	Rafael Aguilar Guajardo
Nuevo Laredo	Tamaulipas	
San Luis Río Colorado	Sonora	Luis Héctor "El Güero" Palma Salazar
Nogales		Emilio Quintero Payán
Hermosillo		
Sinaloa	All the state of Sinaloa	Ismael "El Mayo" Zambada and Baltazar Díaz Vera "El Bala"
Ability to operate in all plazas and to function as liaisons between leaders.		Manuel Beltrán Félix, Rigoberto Campos and Javier Caro Payán

Source: the authors from Astorga (2005) and Valdés (2013)
Note: It is suspected that this distribution of the plazas was actually made by Juan José Esparragoza alias "El Azul" from Sinaloa

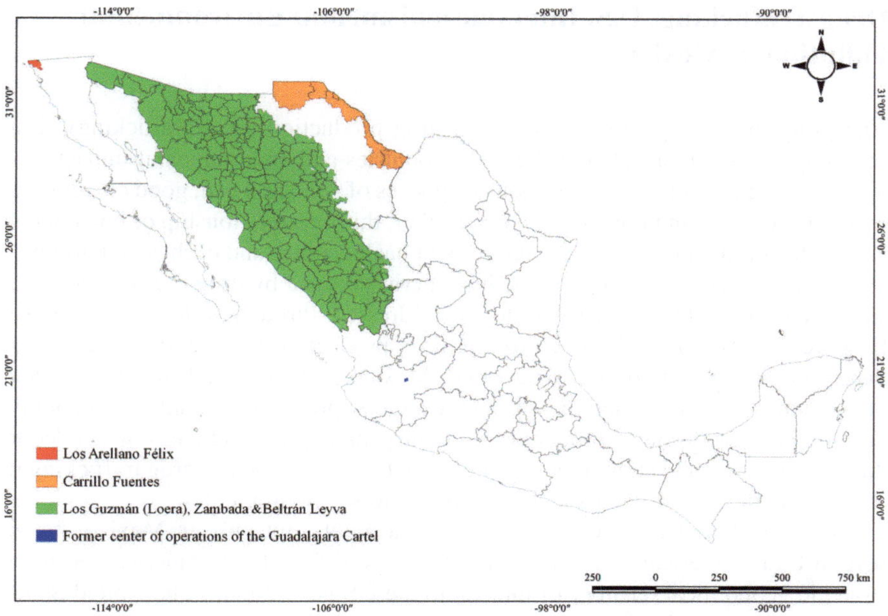

Fig. 6.4 Geographic distribution in the late 1980 of drug trafficking plazas by family clans derived from the absence of the main leaders of the Guadalajara Cartel. *Source*: the authors from Astorga (2005) and Valdés (2013). *Note*: The map shows the location of the plazas of each family clan, but not the territorial extent of each family clan's dominance because the Guadalajara Cartel was the only cartel operating nationwide

- The Arellano Félix family in Tijuana, Baja California
- The Carrillo Fuentes family in Ciudad Juárez, Chihuahua and
- The Guzmán (Loera) family, together with the Zambada and Beltrán Leyva families in Sinaloa, Sonora, Durango, and the Chihuahua mountains

Meanwhile, other local leaders who were not part of any family, such as the Aguilar Guajardo family, "El Güero" Palma, Rafael Chao, and Emilio Quintero, were unable to establish themselves as leading drug trafficking organisations.

Furthermore, the disappearance of this unique organisation (the Guadalajara Cartel) facilitated the entry of new organisations into the illegal drug market, such as García Abrego's in Tamaulipas (the nascent Gulf Cartel), Los Amezcua in Colima, and Los Valencia in Michoacán, with the emergence of new drugs such as amphetamines, mainly led by the latter two groups.

In summary, drug trafficking in Mexico went from monopoly to open competition, and from competition to war for control of territories and new illegal markets. Furthermore, at the end of the 1980s, the two main players in Mexican drug trafficking disappeared: the Sinaloa Organisation and the Mexican government's Federal Security Agency, which provided protection for the Sinaloa Organisation's drug trafficking activities.

Drug Trafficking, U.S. Interventionism, and Environmental Pollution in Mexico

Analysing the interrelationship between drug production (drug trafficking), drug consumption, bilateral relations between countries and environmental impact cannot be reduced simply to dichotomous responses of good and bad, good citizens and criminals, corrupt and non-corrupt, etc. Rather, this interrelationship of phenomena is multifactorial and involves the interests of each country and each social group.

On the American side, the public health crisis caused by drug use represented a significant problem that needed to be addressed through a federal response. However, efforts focused on controlling drug production in other countries rather than domestic consumption. Meanwhile, Mexico, with its weak institutions and phenomena such as corruption, was aggravated by precarious social and economic development conditions throughout the country, mainly in rural areas, which led to the consolidation and strengthening of criminal activities such as drug trafficking in response to the government's lack of attention to social problems.

However, American interventionism was a reality not only in Mexico, but in several Latin American countries. During the Cold War, Latin America became a strategic geopolitical territory of interest to the US government. The United States "supported" the Colombian state with funds and military training to combat the Medellín and Cali cartels, in countries such as Honduras and Panama, under the discourse of the fight against drug trafficking, the United States had a military and intelligence presence and even covered up the illegal activities of political allies with the participation of the CIA.

According to historical records, individuals such as Matta Ballesteros (who was the link between the Sinaloa Organisation and the Colombian cartels) played an important role in the operation set up by the United States government to finance "rebels" in Nicaragua who were fighting to destabilise the Sandinista government, even with resources obtained from drug trafficking. In this way, the CIA hired Matta Ballesteros' planes to traffic drugs to the United States and return with firearms for the Nicaraguan rebels (see Valdés, 2013, p. 192).

Meanwhile, in Mexico, the United States exerted political pressure on the government to overthrow drug trafficking leaders, destroy drug crops, and dismantle criminal organisations involved in drug production and trafficking within the country. However, although American intervention in Mexican territory may have seemed ideal, it caused waves of unpredictable violence and political, economic, and social instability, as well as environmental harms.

Emergence of (Pre)green Criminology, the Fight Against Drug Trafficking and US Interventionism

During the fight against drug trafficking in Latin America led by the United States, while countries struggled with drug trafficking, impunity, corruption, political instability, lack of development, and left-wing political ideologies, environmental issues did not feature on the national or international political agenda. However, the environmental impact of this conflict soon became evident in the field of criminology. Edwin Sutherland's (1961) work "White Collar Crime" enabled Venezuelan criminologist Rosa del Olmo to analyse this conflict from a critical criminological perspective in the Latin American context.

A group of Latin American academics initiated the project in "White Collar Crime and Latin America" in 1977 with the aim of understanding the relationships between political and economic power in Latin America and how this determined the institutional and legislative structure that removes the crimes of the powerful from the criminal and penitentiary spheres, as well as examining the purely symbolic nature of the law, highlighting important sub-projects on environmental crimes and harms in Mexico, Panama, and Venezuela (see Aniyar de Castro, 1987).

After this academic criminological movement, the first criminological studies on environmental harms in Latin America were carried out (before formally emerging as green criminology in 1990), which Goyes and South (2017) refer to as pre-green criminology. This context allows to understand that, as early as the 1970s and 1980s, the war on drugs was not being waged on US territory (although domestic demand for drugs fuelled drug trafficking in other countries) but in Latin American countries, with major social, economic, and environmental repercussions.

Although the responsibility of each country in the fight against drug trafficking in its territories is not minimised, studies by del Olmo (1987) showed that the United States' fight against drugs focused on interdiction and eradication, through fumigation with toxic chemicals such as Paraquat and Glyphosate (produced by Monsanto) to eliminate marijuana, poppy, and coca plants, but used indiscriminately, affecting food crops and causing food crises in the regions, even with full knowledge that Paraquat causes lung damage and that Paraquat and Glyphosate are lethal to wildlife and marine life.

Paraquat (paraquat dichloride) is a highly toxic herbicide, restricted in use and which can only be applied under certification; it is a powerful contact herbicide, non-selective and broad spectrum, used to control weeds in various crops as a defoliant. Ingestion of paraquat can be fatal and contact with the skin or eyes can cause serious effects. In terms of environmental harms, paraquat is a potential risk to mammals, birds, terrestrial invertebrates, terrestrial plants, and algae (United States Environmental Protection Agency [EPA], 2024).

Glyphosate is a widely used herbicide for controlling broadleaf weeds and grasses that has been used in the United States since 1974, although the EPA (2025) currently considers that there are no risks to human health as long as exposure to glyphosate is minimal or within standards; and recognises that ecological risks are potential for terrestrial and aquatic plants, mammals and birds, and low toxicity for honey bees. However, the International Agency for Research on Cancer (IARC) (2018) has documented that glyphosate, along with other organophosphates such as diazinon, malathion, parathion, and tetrachlorvinphos, is considered "probably carcinogenic to humans (Group 2A)".

The use of these herbicides constituted a transnational crime by the United States, with characteristics of ecobiocide, under the pretext of preventing another crime (del Olmo, 1987) and aggravated by the socio-political conditions of Latin American countries where agrochemical regulations were limited and, due to political pressure and "bilateral collaboration commitments" with the United States in the fight against drug trafficking, they were used indiscriminately, causing environmental harms in countries such as Argentina, Bolivia, Brazil, Colombia, Ecuador, Mexico, and Peru, causing environmental and human victims (Goyes and South, 2017).

These studies highlighted the unequal relationship between Latin American countries and the United States, where each country's drug trafficking policies depended on US needs, even overlooking the environmental, social, and economic harms in their own regions that characterised the 1980s. Although American intervention in Latin American countries was particularly intense during those decades (1970–1980) due to the socio-historical context of this region of the continent, current policies in Latin American countries continue to depend on US needs in terms of narcotics (both in terms of drug demand and trafficking control).

Drug Trafficking During the 1990s: Conflicts, Expansion, and Violence

After the collapse of the Guadalajara Cartel, the 1990s were characterised by direct competition between drug trafficking groups for strategic territories (including protected areas) and new drug markets with increasing demand. According to Valdés (2013), three factors determined the context of drug trafficking in Mexico during this decade and the subsequent 15 years:

1. The United States was the main consumer of drugs, generating billions of dollars in profits for those who could meet the demand.
2. After the collapse of the Guadalajara Cartel's leadership and without central control, the market opened up to other new organisations with the capacity to enter the drug business and
3. With the disappearance of the Federal Security Agency (a government entity), the political barriers that limited (regulated) the emergence of other organisations outside the Guadalajara Cartel also disappeared.

The drug trafficking scene in the 1990s was dominated by criminal organisations such as the Arellano Félix in Tijuana, Amado Carrillo Fuentes in Ciudad Juárez, and the Guzmán Loera and Zambada in Sinaloa and Sonora (see Fig. 6.3). These three organisations, which belonged to the Guadalajara Cartel, became known as the Tijuana Cartel, the Juárez Cartel, and the Sinaloa (Pacific) Cartel.

During this decade, these "new" drug cartels originating from the Sinaloa Organisation consolidated their position as "the great drug cartels in Mexico" and occupied a large part of the drug production and trafficking scene to the United States, resisting and adapting to the new political, economic, and market scenarios of the decade and expanding their territories of control for drug production and trafficking.

The Tijuana Cartel

Located in the city of Tijuana, Baja California, on Mexico's north-western border, the Tijuana Cartel, led by the Arellano Félix family, established itself as one of the most prominent groups in the Mexican drug trade during the 1990s. During the time when the Guadalajara Cartel led all drug trafficking in Mexico, Tijuana played a fundamental role in the transport of drugs to the United States, mainly to California, Nevada, Oregon, and Washington, highlighting the role of the state of California as one of the main consumers of drugs in the United States during this decade.

The Arellano Félix family inherited this territory due to their negotiating capabilities and territorial control in this region of Mexico, and after the collapse of the Guadalajara Cartel, they consolidated their position as the Tijuana Cartel. The cartel's operations consisted mainly of collecting protection money (transit fees), because this region of the country is mainly desert. The cartel did not own drug production territory or large tracts of land that would allow for the agricultural production of poppy and marijuana, as in Sinaloa and Chihuahua. Therefore, its main source of income was charging for the passage of drugs through its territory to the United States.

The main operational strengths of the Tijuana Cartel after the decline of the Guadalajara Cartel were:

1. They secured reliable suppliers of cocaine, marijuana, heroin, and methamphetamines at the national and international levels.
2. They organised a secure transportation system for drug trafficking with police protection to ensure that drug shipments would reach Tijuana from different parts of Mexico and the world.
3. They established an effective force and information and intelligence system that allowed them to control their territories and trafficking routes, including protected areas.

The drug trafficking routes operated by the Tijuana Cartel were centred on the Pacific Ocean region. In the case of cocaine from Colombia, it was transported by boat by Colombians to the coast of Oaxaca, where it was received by Mexicans and then transported by sea to the coast of Michoacán. From Michoacán, it was transported by air to Ensenada in Baja California, and the final route from Ensenada to Tijuana was by sea to La Bufadora beach in Tijuana, where it was safeguarded by police and stored in warehouses (Fig. 6.5).

Even with the collapse of the Colombian cartels (Cali Cartel and Medellín Cartel), the Tijuana Cartel found cocaine suppliers in the Revolutionary Armed Forces of Colombia (FARC) as early as 1998. At first, the Tijuana Cartel paid the Colombian guerrillas in cash and later with firearms that were trafficked by members of the Mexican Federal Highway Police (Valdés, 2013).

Nevertheless, the Tijuana Cartel did not only traffic cocaine; the entry of methamphetamines into the drug trade allowed this cartel to import and traffic methamphetamines and heroin from Asia. According to historical records, methamphetamines were brought in from Europe and arrived at Mexico City airport, from where they were also trafficked by air to Mexicali in Baja California, as the Mexicali Plaza during the time of the Sinaloa Organisation passed to the Arellano Félix family after the arrest of the plaza's leaders. Mexicali was also used by the Sinaloa Cartel led by "El Chapo" Guzmán and "El Güero" Palma (head of the San Luis Río Colorado Plaza in Sonora) upon payment of transport fees.

The Tijuana Cartel consolidated alliances with other smaller drug traffickers who needed to cross drugs across the border; for example, the Amezcua Contreras family from Guadalajara were the main producers of methamphetamine in Mexico, with laboratories in Colima, Jalisco, Michoacán, and Mexico City, and formed alliances

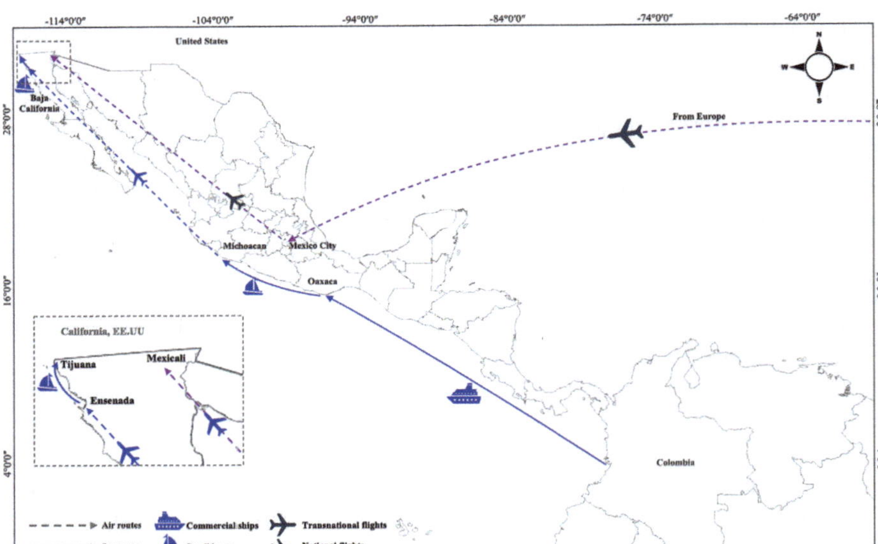

Fig. 6.5 Main drug trafficking routes of the Tijuana Cartel during 1990. *Source*: the authors with information from Valdés (2013)

with the Tijuana Cartel to smuggle drugs across the country's north-western border. In addition, the Arellano Félix family established relationships with criminal networks involved in drug trafficking in Japan and Hong Kong, China, to bring heroin into Mexico, which was paid for with cocaine that was trafficked back to Asian countries.

According to historical records, the Tijuana Cartel established itself as one of the longest-running and most sustainable cartels in Mexico. It is even argued that for 20 years, the Arellano Félix brothers were never captured or imprisoned, either in Mexico or in the United States. In addition, the cartel's payroll included prosecutors, police commanders, delegates, and other public servants, including US border and security agents, through one of the main practices that characterised the Tijuana Cartel: bribery and expensive gifts.

Mexican government protectionism, although on a smaller scale than with the Sinaloa Organisation, allowed the Tijuana Cartel to consolidate and strengthen its criminal networks and territories of control. Through heavy-handed control and violence, they established themselves on Mexico's north-eastern border and gained total control of the maritime and air routes used for drug trafficking.

The Juarez Cartel

Located in the northern border region of the state of Chihuahua, Ciudad Juárez became the centre of the Ojinaga Plaza after the death of Pablo Acosta and the arrest of Aguilar Guajardo, the former owner of the plaza, and following the collapse of the Guadalajara Cartel, Amado Carrillo Fuentes, with the support of his two brothers, became the leader of the Ciudad Juárez Cartel or Juarez Cartel.

The geographical location of the Ciudad Juárez Cartel under Carrillo Fuentes' command was strategic (see Fig. 6.3), allowing drugs to be brought directly into the centre of the United States and the Chicago region. During this time, the Ciudad Juárez Cartel consolidated cocaine trafficking routes from Colombia with government protection, mainly from police officers and commanders whom it paid hundreds of thousands of dollars for protection.

Furthermore, due to the distance between Chihuahua's northern border and central Mexico, Carrillo Fuentes had an aviation company in Torreón, Coahuila, since the late 1980s, which transported more cocaine than people, earning him the alias "El Señor de los Cielos" (The Lord of the Skies). Unlike the Tijuana Cartel, Carrillo Fuentes provided the Juárez Cartel with broader government protection, not only from local police and commanders, but also from the military and representatives of federal agencies who dictated and formulated police operations against drugs from Mexico City.

Carrillo Fuentes positioned the Juárez Cartel as the centre of operations for cocaine trafficking from different regions of Mexico. It operated from Cuernavaca and Ciudad Juárez, but the drugs came from other places such as Cancún, Guadalajara, Torreón, and Hermosillo, transported mainly by small aircraft with half-tonne shipments of cocaine per trip (Fig. 6.6). According to records, it is estimated that the Juárez Cartel transported at least 60 tonnes of cocaine to the United States per year, approximately 120 flights each year, one every third day (Valdés, 2013).

Unlike other criminal associations such as the Tijuana Cartel, which are characterised by violence, the Juárez Cartel was initially characterised by a conciliatory attitude and good relations with other criminal groups and even more so with government authorities, which allowed this cartel to remain operational under a certain degree of government protection and in association with the leaders of Sinaloa and with a network of espionage towards other cartels. Despite the determination to eliminate the Arellano Félix from the drug market scenario, this period was characterised by violence due to the struggle to consolidate a single drug association, such as the Sinaloa Organisation.

The Sinaloa (Pacific) Cartel

After the collapse of the Guadalajara Cartel, Sinaloa continued to maintain its power in drug production (in the Golden Triangle) and trafficking, but with some changes. The state of Sinaloa does not border the United States, so the drugs produced

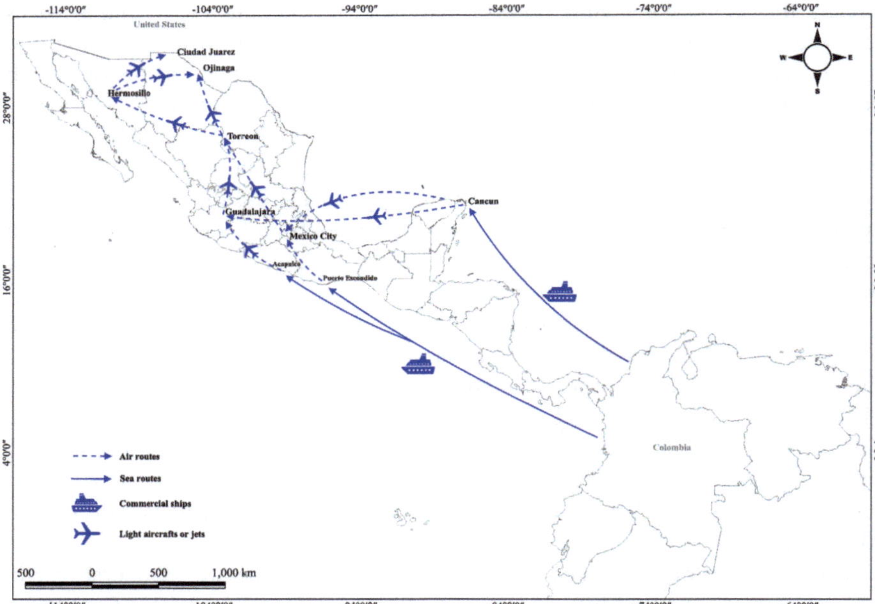

Fig. 6.6 Main drug trafficking routes of the Juarez Cartel during 1990. *Source*: the authors with information from Valdés (2013) and InSight Crime (2020)

(marijuana and poppy gum) had to be moved to the border, and to achieve this, alliances were established with other criminal groups along the border. Ismael "El Mayo" Zambada led this organisation, and his main allies were Joaquín "El Chapo" Guzmán Loera and Héctor "El Güero" Palma Salazar, who controlled Tecate, Baja California, and San Luis Río Colorado, respectively. They later seized power in Mexicali, and it is suspected that Juan José Esparragoza "El Azul" was also part of this alliance, linking them to Amado Carrillo Fuentes of the Juárez Cartel.

This association became the new drug trafficking force based in the state of Sinaloa, occupying the border strip of the state of Sonora. During the evolution of this organisation in the 1990s and early 2000s, El Chapo became the most prominent figure in this association. In collaboration with his friend El Mayo, they controlled much of the drug trafficking in north-eastern Mexico. El Chapo had his base of operations in Guadalajara, Jalisco, and Agua Prieta, Sonora. They also had safe houses in Culiacán, Mexicali, and Tecate, as well as drug production ranches (marijuana and poppy) in Chihuahua, Sinaloa, Sonora, and Durango (Fig. 6.7).

The Sinaloa cartel, although historical records credit El Chapo with most drug trafficking activities, took advantage of the border between Sonora and Arizona, managing to cross drugs using various techniques such as small aeroplanes, they smuggled drugs across the border in tourists' suitcases on commercial flights, large ship containers, speedboats, semi-submersible boats, and catapults that launched 50 kg packages of marijuana over the border fence.

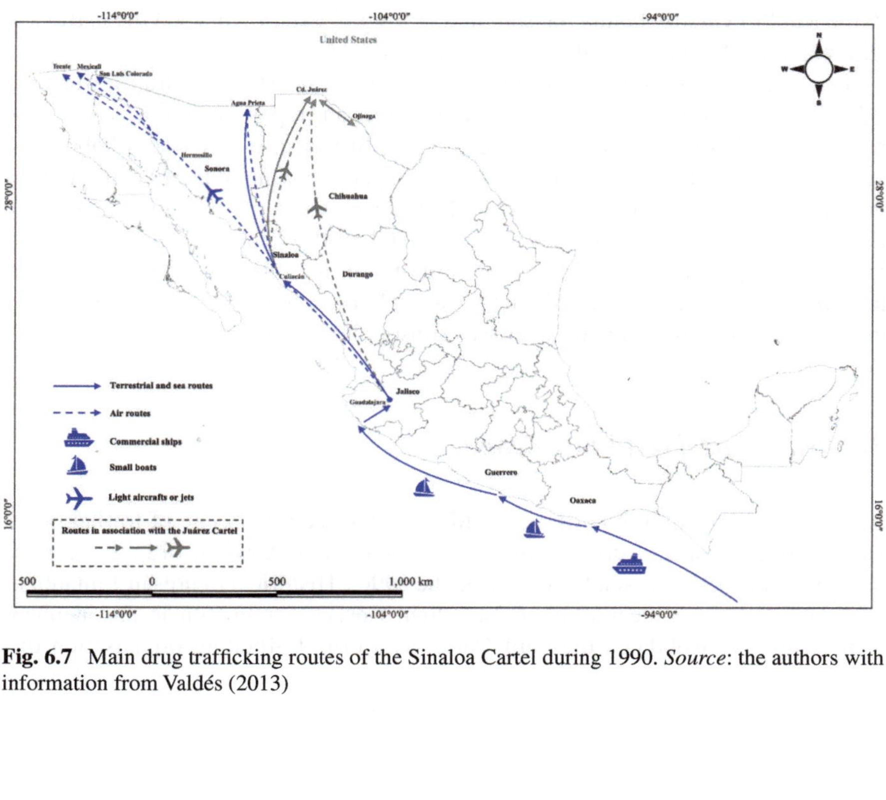

Fig. 6.7 Main drug trafficking routes of the Sinaloa Cartel during 1990. *Source*: the authors with information from Valdés (2013)

The territorial reach of the Sinaloa Cartel extended as far as the Canadian border in the United States. During this decade, there was even a report of an automated irrigation system for growing marijuana in Nicolet National Forest in the US state of Wisconsin, guarded by Mexicans with AK-47 rifles, which supplied marijuana to the Chicago area (Valdéz, 2013; Actualidad RT, 2010). Furthermore, with the incursion of methamphetamines into the Mexican drug market, the production of this new drug was centralised in Guadalajara, Jalisco, with the creation of narco-laboratories in the city. The trafficking of this new drug was incorporated into the marijuana packages that travelled along the already established routes.

Among the historical events that marked the Sinaloa Cartel's operations were the struggles for drug trafficking territories and routes between the Tijuana Cartel, led by the Arellano Félix brothers, and the association between the Sinaloa Cartel and the Juárez Cartel. The violent clashes between the cartels and their espionage networks led to a shootout on 24 May 1993 at Guadalajara airport, led by sicarios working for the Arellano Félix who arrived from San Diego, California, with the aim of killing El Chapo. During the confrontation, Catholic Cardinal Juan José Posadas Ocampo was killed by 14 gunshots while picking up Apostolic Nuncio Girolamo Priggione at the airport.

As a result of this scandalous event, the Mexican government, in collaboration with the Guatemalan army, arrested Joaquín Guzmán Loera in Guatemalan territory and extradited him to Mexico, where he was sentenced to 20 years in prison, while Rafael Arellano Félix was arrested and imprisoned in December of that same year. In this new scenario, the Sinaloa Cartel was not left leaderless, as "El Mayo" Zambada, "El Güero" Palma Salazar, and "El Azul" Esparragoza led the cartel and still maintained alliances with the Juárez Cartel.

This decade saw important events in the context of Mexican drug trafficking. With El Chapo in prison, Amado Carrillo played an important role in association with the Sinaloa Cartel. Although there are no historical records of collaboration agreements between the two cartels, they managed to maintain their position in the drug market. With the arrest of Héctor "El Güero" Palma in 1995 and the death of El Señor de los Cielos, leader of the Juárez Cartel, in 1997, the challenges to remaining in the drug market were many. Furthermore, during this decade, one of the cartels most representative of the Mexican War on Drugs and the violence generated by the clashes in the following decades was consolidated: the Gulf Cartel.

The Gulf Cartel (and Los Zetas)

Although historically the northeast of Mexico has been the scene of trafficking in various illegal products, drug trafficking appeared on the Mexican drug scene in the late 1980s and became established during the 1990s. Historical figures in Tamaulipas such as Juan Nepomuceno Guerra and Juan García Ábrego (uncle and nephew, respectively) consolidated the Gulf Cartel as a drug trafficking organisation that mainly trafficked cocaine to Texas and from there to the east coast of the United States.

Records show that in early 1990, Juan Nepomuceno Guerra was arrested, and in 1996, Juan García Ábrego was arrested. The actions of both men allowed the Gulf Cartel to consolidate its power under the protection of the Mexican police and government in the states of Tamaulipas and Nuevo León. Unlike other drug cartels during this period, the Gulf Cartel was not led by family clans, which would later lead to conflicts among its members over the leadership of the organisation.

After the imprisonment of Juan García Ábrego and following adjustments within the organisation, Osiel Cárdenas Guillén headed the Gulf Cartel, establishing trafficking routes, purchasing government protection, organising cross-border transfers, and subjugating other small drug trafficking groups in the region.

Unlike the plaza system operated by the former Sinaloa Organisation, the Gulf Cartel operated throughout the state of Tamaulipas with three major border crossings: Nuevo Laredo, Reynosa, Matamoros and other smaller border settlements, covering a territorial extension of 370 km with the state of Texas, unlike the plazas of other drug cartels, which controlled small towns or in some cases only one town, for example, Tijuana.

The main drug trafficked by the Gulf Cartel was cocaine, which was brought in from Guatemala and arrived by sea to the coasts of the states of Campeche and Veracruz. It was then transported by land to the territory of Tamaulipas, but it also arrived by sea directly to the port of Tampico-Altamira. In addition, when it entered by land, it traced a route from Guatemala through the states of Chiapas, Campeche, Tabasco, and Veracruz (Fig. 6.8).

Subsequently, the Gulf Cartel opened a new route known as Pacífico-Tamaulipas, which began in the states of Guerrero and Michoacán, crossing through San Luis Potosí and Zacatecas, to Torreón and Saltillo in the state of Coahuila and Monterrey in the state of Nuevo León, and from there to any of the border points in Tamaulipas. These routes show that drug trafficking in Mexico during this decade covered the entire national territory.

One of the most significant events within the Gulf Cartel organisation was the creation of Los Zetas as a personal guard for Osiel Cárdenas Guillén, the cartel's main leader. This personal guard was the leader's response to a group of criminal drug bands that had been left uncontrolled since the capture of García Ábrego in 1996, which were attempting to take over other illegal markets and territories within the state of Tamaulipas. These other smaller drug trafficking groups also hired personal guards for their protection. For example, Eduardo Costilla, alias "El Cos" and Víctor Manuel Vázquez, alias "El Meme" hired two gangs known as Los Sierras and Los Tangos, while Ezequiel Cárdenas Guillén (Osiel's brother) armed a group called Los Escorpiones.

Osiel Cárdenas Guillén wanted an elite group with expertise in protection and weaponry, so he hired deserters from the Mexican Army. In a short time, more than 50 members of this elite army unit and other areas formed Osiel Cárdenas' praetorian guard: Los Zetas. By the beginning of 2000, the number of members in Los Zetas had risen to more than 300. In addition, they recruited members of the

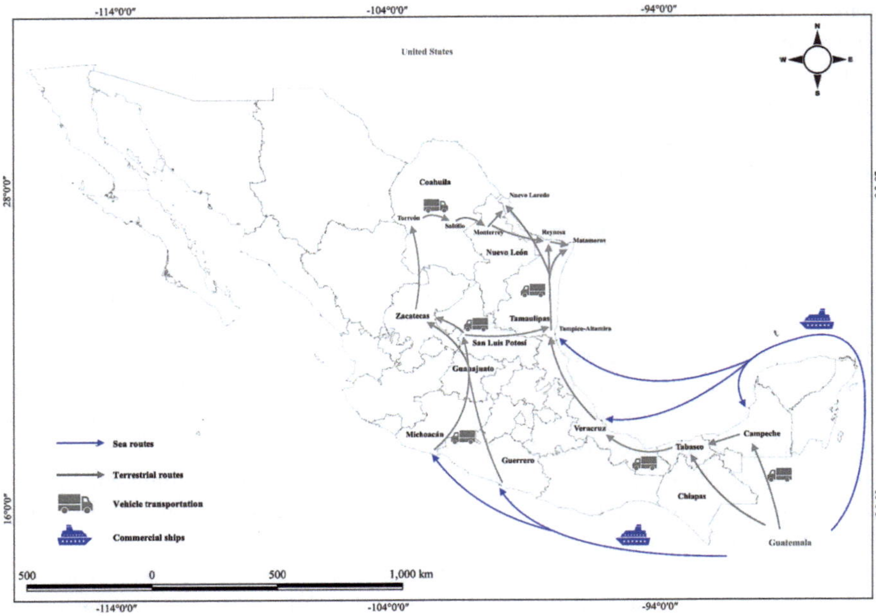

Fig. 6.8 Main drug trafficking routes of the Gulf Cartel during 1990. *Source*: the authors with information from Valdés (2013)

Guatemalan army known as "kaibiles" to join Los Zetas in Guatemala. These elite soldiers were involved in the ethnic cleansing of indigenous populations in Guatemala during the 1980s and were known for their savage and ruthless methods of combat and confrontation.

This led to the creation of training camps in states such as Tamaulipas, Guerrero, and Veracruz for new sicarios trained in all kinds of skills by American and Israeli military personnel in jungle and forest areas, often in protected areas. According to Grillo (2011), this situation was particularly characteristic of the Gulf Cartel and its armed faction, Los Zetas, so that it was no longer just corruption, but traitors to the country who deserted the army to join the forces of drug trafficking in Mexico.

The Zetas enabled the expansion of the Gulf Cartel and its positioning among the major drug cartels during the 1990s and early 2000s. The former sicarios of the other cartels were no match for the military training of the Zetas in the Gulf Cartel, which allowed them to occupy more territories such as Michoacán and open a new drug trafficking route beyond the Gulf of Mexico. In addition, they established trafficking routes for other drugs such as methamphetamines and amphetamines produced in the state of Michoacán, which were transported via the cartel's routes. Furthermore, the raw materials to produce methamphetamines and amphetamines arrived directly at the Port of Lázaro Cárdenas in Michoacán from Asia.

In subsequent decades, Los Zetas occupied a large part of the drug trafficking scene in Mexico, not only as the armed faction of the Gulf Cartel, but also as an organisation involved in illegal activities such as kidnapping, extortion, racketeering of small and medium-sized businesses, protection fees, contract killings, and other activities related to drug trafficking. They carried out these activities with extreme violence, affecting not only government entities but also society.

"Minor" Drug Trafficking Groups

This expansionist movement by the Gulf Cartel and its armed gang, Los Zetas, led to the occupation of locations and territories belonging to other smaller drug producers in the country. As a result, although historical records show the existence of criminal groups such as Los Valencia and the Milenio Cartel in the state of Michoacán, which had criminal associations with the Juárez Cartel, the Tijuana Cartel and the Sinaloa Cartel were unable to resist the violent onslaught of Los Zetas to dominate the state.

The Los Valencia criminal group operated quietly but sophisticatedly, transporting cocaine from Colombia to the coast of Michoacán state via tuna boats and then moving it by land and air, depending on the purchasing cartel. Unlike other cartels, Los Valencia worked quietly and without flaunting their wealth. Even the Mexican government was unaware of their existence until the DEA, in collaboration with the Colombian government and later with the Mexican government, dismantled the organisation. Although they were able to remain operational after the government intervention, they were powerless against Los Zetas.

Later, in this geographical area of the state of Michoacán, with Los Zetas in total control at the beginning of 2000, a new independence movement emerged within Los Zetas. The constant abuses against society by this criminal group led members of this criminal association and members of society in Michoacán to form a new independent group to confront the various criminal cells of Los Zetas. This movement was called La Familia (Michoacana), and this group had three main characteristics: a clear ideology about the functioning of the organisation with almost religious ideas, a left-wing political agenda, and a demand for control over the territory.

La Familia Michoacana, like Los Zetas, established new modalities of operation for criminal groups involved in drug trafficking in Mexico. Los Zetas, on the one hand, with their use of extreme violence, military training, and attacks on civil society, inherited similar characteristics to La Familia Michoacana. This new group had a criminal style with a strategy that differed from all other organised crime and drug trafficking groups in Mexico at that time.

According to records, La Familia Michoacana operated under the following assumptions: a) a discourse that was publicised with the idea of being an organisation created by society to defend itself from foreign criminals and bad governments that oppress it, and b) an attempt to hide its criminal nature behind a supposed social welfare and support base, built through the distribution of social welfare centres, such as drug rehabilitation centres, businesses with local entrepreneurs, and public works projects carried out by local councils controlled by this criminal association (Valdés, 2013).

This new criminal group used violence to establish itself in the state of Michoacán, expelling and murdering Los Zetas. It also spread to central Mexican states such as the State of Mexico and Mexico City, using practices very similar to Los Zetas (extortion, protection rackets, etc.). However, among its most significant activities was its imposition on the governments of the state of Michoacán, such that every town council or municipality within the state was administered and coordinated by La Familia. They also implemented strategies of corruption and infiltration into government positions to divert state economic resources to the bank accounts of this criminal group, mainly through resources for public infrastructure projects in the state.

This drug trafficking scenario during the 1990s shaped the landscape of drug trafficking in the following decades. The presence of drug trafficking groups and other criminal activities in large areas of territory often coincided with protected areas, which were used as territories for planting, harvesting, and transporting drugs, but also as training zones for sicarios and later for the creation of drug laboratories.

Conclusions

After the collapse of the Guadalajara Cartel in the 1990s, the emergence and consolidation of other criminal organisations linked to drug trafficking were imminent. During this decade, drug trafficking routes were consolidated, leaders emerged and succumbed, but a scenario was also established that determined the conditions under which drug trafficking operated in Mexico for the following decades.

In this context, wilderness areas were used as fields for growing poppy and marijuana, as well as territories for drug trafficking. These territories were and have been coveted by drug traffickers and have played a major role in causing conflicts between criminal groups.

At the end of the 1990s, there were 197 marine and terrestrial areas designated for environmental conservation in Mexico. Diachronically, until the end of the 1990s, there was no national institution to administer these areas, and as a result, the government neglected their management because these areas existed only on paper. Furthermore, it should be noted that during this decade, government efforts to control and eliminate drug trafficking, and in particular drug production by the United States government, led to serious cases of environmental contamination by herbicides, but also by agrochemicals used for drug production by drug trafficking groups.

The emergence of new drug cartels during this decade influenced the operating methods of the newly formed National Commission for Protected Natural Areas (CONANP) in 2000 (CONANP, 2022), because environmental conservation and development actions within and near protected areas were determined by the presence of drug cartels in those territories, and activities such as the drugs produced (mainly marijuana and poppy), forms of violence, means of drug trafficking, as well as their relationships with Mexican and foreign government entities.

There is a lack of historical information on the presence of criminal groups involved in drug trafficking in conservation areas during the decades prior to 2000. Therefore, studying this scenario of drug trafficking in Mexico's protected areas is essential for three reasons: a) to document the environmental impact of drug production practices and the government's (national or foreign) fight against drug production, b) to understand the forms of environmental conservation implemented in protected areas during the first decades of the emergence of CONANP as the government institution in charge of managing protected areas, and c) to understand, from a historical perspective, the evolution of drug trafficking in Mexico, considering the importance of protected areas as strategic territories for drug production.

References

Actualidad RT. (2010). *Descubren sembrados de marihuana en un parque nacional en Wisconsin*. PeriódicoActualidadRT(13deagostode2010).Availableonhttps://actualidad.rt.com/actualidad/view/15417-Descubren-sembrados-de-marihuana-en-un-parque-nacional-en-Wisconsin

Aguilar, H. (2009). *Narco. Historias extraordinarias*. Nexos (mayo 2009). https://www.nexos.com.mx/?p=12886

Aniyar de Castro, L. (1987). *Criminología de la liberación*. Editorial de la Universidad de Zulia.

Astorga, L. (2005). *El siglo de las drogas. El narcotráfico, del Porfiriato al nuevo milenio*. Plaza y Janés.

Bailey, J., & Godson, R. (2000). *Organized crime and democratic governability: Mexico and the U.S.-Mexican borderlands*. University of Pittsburgh Press. https://doi.org/10.2307/j.ctt9qh7d4

Baum, D. (1996). *Smoke and mirrors. The war on drugs and the politics of failures*. Little, Brown and Company.

Carpio-Domínguez, J. (2021). Crimen Organizado (narcotráfico) y conservación ambiental: el tema pendiente de la seguridad pública en México. *Revista CS*, 33, 237-274. https://doi.org/10.18046/recs.i33.4076

Comisión Nacional de Áreas Naturales Protegidas (CONANP). (2022). *Sistema Nacional de Áreas Protegidas (SINAP)*. Gobierno de México. SEMARNAT. https://www.gob.mx/conanp/acciones-y-programas/sistema-nacional-de-areas-protegidas-sinap

del Olmo, R. (1987). Aerobiology and the war on drugs: A transnational crime. *Crime and Social Justice, 30*, 28–44. https://www.ojp.gov/ncjrs/virtual-library/abstracts/aerobiology-and-war-drugs-transnational-crime

El Heraldo de Chihuahua. (2021). *=Qué pasó con el Rancho El Búfalo de Caro Quintero? aquí los detalles de la historia*. El Heraldo de México (May 5, 2021). Available on https://heraldodemexico.com.mx/tendencias/2021/5/5/que-paso-con-el-rancho-el-bufalo-del-caro-quintero-aqui-los-detalles-de-la-historia-292582.html

Goyes, D., & South, N. (2017). Green criminology before "green criminology": Amnesia and absences. *Critical Criminology, 25*, 165–181. https://doi.org/10.1007/s10612-017-9357-8

Grillo, I. (2011). *El Narco: inside Mexico's criminal insurgence*. Bloomsbury Press.

InSight Crime. (2020). *Cartel de Juárez*. InSight Crime Analysis Investigation of Organized Crime. https://insightcrime.org/es/noticias-crimen-organizado-mexico/cartel-de-juarez-perfil/

International Agency for Research on Cancer (IARC). (2018). *IARC monograph on glyphosate*. World Health Organization. https://www.iarc.who.int/featured-news/media-centre-iarc-news-glyphosate/

Melo, O. (1998). Narcotráfico y democracia: La experiencia colombiana. In C. Malamud & E. Joyce (Eds.), *Latin American and the multinational drug trade* (pp. 63–96). St. Martin's Press - Palgrave Macmillan.

Morales, L. (2017). *La paz y la protección ambiental en Colombia: propuestas para un desarrollo rural sostenible*. El diálogo, propuesta para las Américas. https://www.thedialogue.org/wp-content/uploads/2017/01/Envt-Colombia-Esp_Web-Res_Final-for-web.pdf

Rosas, E. (2025). *La Guerra contra las Drogas: el Caballo de Troya del Complejo Militar Americano*. Nexus (mayo 14, 2025). https://seguridad.nexos.com.mx/la-guerra-contra-las-drogas-el-caballo-de-troya-del-complejo-militar-americano/

Salazar-Pérez, R., & Fernández-Velázquez, J. (2016). Geopolítica del narcotráfico. Un escenario para América Latina desde México. *Veredas, 32*, 219–244. https://veredasojs.xoc.uam.mx/index.php/veredas/article/view/402/397

Santana, A. (2018). Elecciones y crisis en Honduras (2017-2018). *CariCen, Revista de Análisis y Debates sobre el Caribe y Centroamérica, 7*, 4–14. https://investigacion.politicas.unam.mx/caricen/flips/flip_7/caricen7.pdf#page=5

Sutherland, E. (1961). *White collar crime*. Holt, Rinehart and Winston.

United States Environmental Protection Agency (EPA). (2024). *Paraquat dichloride*. US Government. https://www.epa.gov/ingredients-used-pesticide-products/paraquat-dichloride

United States Environmental Protection Agency (EPA). (2025). *Glyphosate*. US Government. https://www.epa.gov/ingredients-used-pesticide-products/glyphosate

Valdés, G. (2013). *Historia del narcotráfico en México*. Editorial Aguilar.

Chapter 7
Environmental Conservation and the Prelude to the "Mexican War on Drugs" in Mexico (2000–2006)

Introduction

The drug trafficking scenario during this period was characterised by the diversification of criminal groups engaged in drug trafficking in Mexico. This meant that different criminal groups had different operating methods, which influenced the ways in which environmental conservation activities were conducted in Mexico's protected areas.

During this period, the Comisión Nacional de Areas Naturales Protegidas (CONANP) (National Commission for Protected Natural Areas) was created as a government entity responsible for administering protected areas. This involved the permanent presence of an organisational structure with directors and inspectors for each federally protected area. In this sense, environmental conservation activities were influenced by the presence of drug traffickers in territories designated as protected areas.

Considering that the management programmes[1] for each protected area did not consider the presence of criminal groups, CONANP personnel carried out environmental conservation activities in a specific scenario involving poppy and marijuana crops, the presence of armed men, armed confrontations, and a polarisation between illegal drug production activities and sources of employment for communities near and within protected areas.

This alternation between public security and environmental conservation activities reveals a complex scenario in which political, social, economic, environmental, and public security factors coincided in specific, geographically defined territories that did not occupy a fair place on government agendas or in academic analysis.

---

[1] Management plans for protected areas are technical documents that establish strategies and actions for the conservation, protection, and sustainable use of natural resources within a protected area. These plans serve as guidelines for the management of protected areas, defining objectives, zoning, regulations, and permitted or restricted activities (CONANP, 2025a).

J. L. Carpio-Domínguez et al., *Drug Trafficking and Protected Areas in Mexico*,
https://doi.org/10.1007/978-3-032-18218-0_7

However, it should be emphasised that there are no historical records of the activities of criminal groups involved in drug trafficking within protected areas during this period. Even at the institutional level within CONANP, the experiences of inspectors and directors who were in contact with drug cultivation fields and drug traffickers remain only in the oral records and memories of some inspectors and directors, and even of personnel who no longer work at CONANP.

To address this information gap, data was collected directly from CONANP personnel who conducted environmental conservation activities after the institution's creation and prior to the Mexican War on Drugs (2007–2012). Their experiences were recorded with the aim of understanding and documenting, through primary sources, the context of environmental conservation in territories invaded by criminal drug trafficking groups in Mexico.

Emergence of the National Commission of Natural Protected Areas (CONANP)

In the context of struggles for the expansion of territories controlled by criminal drug trafficking groups throughout the country, the National Commission for Protected Natural Areas (CONANP) was created on 5 June 2000 during the presidency of Ernesto Zedillo Ponce de León, as a decentralised institution of the Secretariat of Environment and Natural Resources (Secretaria de Medio Ambiente y Recursos Naturales) (SEMARNAT).

CONANP was created with the aim of managing Protected Natural Areas[2] and protecting the country's natural heritage. It had a total of 150 Protected Natural Areas, 22 of which were marine areas. The total area under environmental protection in the country was 9,637,825 hectares until that year (Comisión Nacional de Áreas Naturales Protegidas [CONANP], 2022).

Until 2000, all states in the country had protected areas, with Chiapas ($n = 17$, 8.6%) being the state with the most areas designated for environmental conservation, followed by Quintana Roo ($n = 13$, 6.6%), the State of Mexico ($n = 12$, 6.1%), Michoacán ($n = 11$, 5.6%), Mexico City ($n = 10$, 5.1%), Jalisco ($n = 10$, 4.1%), Baja California ($n = 7$, 3.6%), Baja California Sur ($n = 7$, 3.6%), Coahuila ($n = 7$, 3.6%), and Oaxaca ($n = 7$, 3.6%).

During this period, CONANP regionalised protected areas into nine regions (Fig. 7.1) within the national territory with the aim of achieving greater institutional coverage and making conservation activities more effective (CONANP, 2015; González-Ocampo et al., 2014).

[2] In Mexico these areas are named "Natural Protected Areas", but throughout the book are referred to as "protected areas" in order to homologate the term with other countries that have this legal and institutional figure for environmental protection.

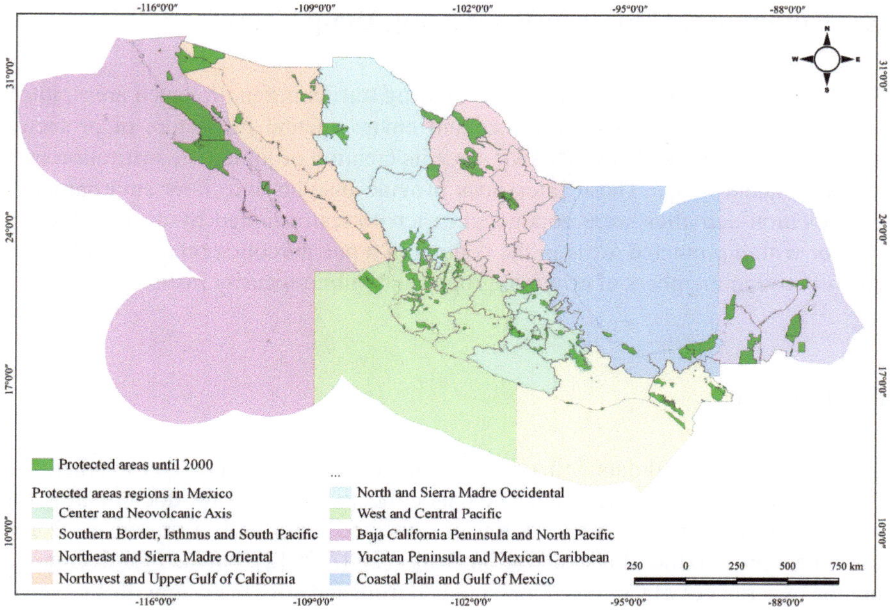

Fig. 7.1 Regions and protected areas in Mexico until 2000. *Source*: The authors with data from CONANP (2025b)

The formalisation of a government institution to manage and administer protected areas led to the formal and institutional implementation of environmental conservation activities, scientific studies, and sustainable development activities in and with rural and indigenous communities near and within protected natural areas.

The hiring of environmental inspectors and a whole range of government jobs to manage each protected natural area made it possible to guide environmental conservation actions locally and address local needs in collaboration with rural and indigenous communities, including verification of compliance with regulations, a controlling access to and use of infrastructure, promoting environmental education and community participation, and monitoring and surveilling natural resources (CONANP, 2018).

The implementation of these activities by the personnel (directors and inspectors) of the recently created CONANP occurred in the context of the expansion and consolidation of drug trafficking groups in Mexico, and the new decrees on protected areas often coincided with territories that had already been occupied by drug traffickers for decades, but also in the historical political context of political change with the departure of the Partido Revolucionario Institucional (PRI) from the presidency and the rise of the Partido Acción Nacional (PAN) to the presidency of Mexico.

Environmental Conservation Among Drug Crops

To understand the influence and impact of drug trafficking in protected areas, interviews were conducted with directors and environmental inspectors of protected areas who have worked at CONANP since its creation or joined the institution over the subsequent years. These interviews provide insights into how environmental conservation activities were conducted in territories occupied by drug traffickers near or within protected areas in the context of a pax narcotica represented by alliances between members of criminal groups and public security institutions.

Methodology

The methodology and data collection instrument were implemented to gather data that is used throughout the book. The data collected is useful for documenting the reality of environmental conservation in protected areas with a presence of drug traffickers prior to the Mexican War on Drugs (2007–2012), but also for documenting the experiences of CONANP personnel during and after the war, with a focus on the perspective from protected areas.

Qualitative sampling was applied, utilising the chain or snowball technique, which involves locating informants who meet the selection criteria and requesting them to refer other potential participants (Bernard, 2006). The sample consisted of 10 directors and 5 environmental inspectors from CONANP in 12 protected areas in Mexico, 6 women, and 10 men.

At the express request of the interviewees, their identities, job positions, names of protected areas, or regional headquarters are kept confidential, as some of the informants contacted did not agree to be interviewed, citing fears for their safety if they provided information about criminal groups within protected areas or in nearby regions. Even so, whenever possible, information will be provided on the dominant criminal group in the region in order to contextualise the group's working methods and its relationship with conservation activities conducted by CONANP personnel.

Data Collection Instrument

The interviews were designed to obtain a diachronic perspective of the phenomenon under study (before, during, and after the Mexican War on Drugs), as well as to obtain subcategories that would allow for a deeper understanding of the challenges of conservation in Mexico. The interview design incorporated four preliminary categories: (1) environmental conservation before the war, (2) conservation strategies during the war, (3) conservation strategies in the post-war period, and (4) drug trafficking and current threats to conservation.

The first category analyses the situation of protected areas prior to the war, the activities conducted for environmental conservation, and the conditions under which these activities were conducted, as well as strategies for implementing conservation and local development activities in communities near or within protected areas.

The second category analyses the context of environmental conservation in protected areas during the Mexican War on Drugs (2007–2012), examining the role played by state forces within the areas, the influence of armed conflict on citizen participation and tourism in the area, and the main strategies used to pursue and guarantee the security–conservation binomial. This category aims to highlight the war scenario within protected areas.

Category three analyses changes in the activity patterns of criminal groups within protected areas after the war and their influence on local ecosystems, as well as the environmental consequences of this period of conflict between criminal groups and the state from the perspective of environmental directors and inspectors. This category aims to understand the effects of the Mexican War on Drugs on these institutions, as well as the strategies used to resume conservation actions.

The fourth and final category aims to identify the main current threats to conservation activities in protected areas, within the context of public safety, arising from the continued presence of criminal groups involved in drug trafficking within protected areas.

Working Peacefully, Each Attending to Their Own Business

Considering that at the time CONANP was created, there were already protected areas designated throughout the national territory, the beginning of environmental conservation activities in these areas presented several challenges for the environmental directors and inspectors hired to manage each area. The boundaries of each protected area are clearly defined and established in the decree creating each protected area. One of the main problems faced by CONANP personnel was the inability to access the entire territory designated as protected.

Environmental inspectors were unable to enter areas within the protected zone due to the high risk to their personal safety posed by conducting conservation activities in areas located several hours or days away from the nearest town, where there were also fields of marijuana and poppy cultivation, as well as established routes for drug trafficking.

Furthermore, the location of many of the protected areas meant that government surveillance did not reach them. Their remoteness from large urban centres and state capitals, combined with their sparse population, made these territories strategic for the planting, harvesting, and trafficking of drugs. Mountainous areas and their valleys, jungles, beaches, and uninhabited islands became valuable territories for drug trafficking (Fig. 7.2).

Fig. 7.2 Orography of protected areas in the northeast and Gulf of Mexico. *Note*: Figures (**a**) and (**b**) correspond to protected areas in the Region of North and Sierra Madre Occidental; figures (**c**) and (**d**) correspond to protected areas in the Region of Coastal Plain and Gulf of Mexico. Photographs: José Luis Carpio-Domínguez

However, environmental conservation actions in the years prior to the Mexican War on Drugs (2007–2012) were implemented without major problems. There was a tacit and binding agreement whereby environmental directors and inspectors in protected areas did not interfere in the drug production activities of drug traffickers.

> Since the creation of the protected area, we have been unable to enter areas that are part of the protected area [...] to conduct biological monitoring, surveillance or conservation projects (01-Environmental inspector, Sinaloa Cartel territory).
>
> When I was designated director of the protected area and arrived at the offices, a group of drug producers and their leaders had already been sent by "the boss" to inform me about the drug production situation in the area and to explain my duties and those of the team of inspectors, which at that time consisted of only three people, including myself. Once we talked, we agreed that we would not get involved in their business and that they would support us with whatever equipment and personnel we needed for fighting fires, such as picks, shovels, torches, etc. (03-Director of protected area, Sinaloa Cartel territory).
>
> When we arrived at the protected area, they "warned" us [threats], telling us not to get involved in their business and to close our eyes if we saw anything [drug trafficking], that we could monitor wildlife and activities with the fishing communities, but without interfering in their activities (05-Director of protected area, Tijuana Cartel territory).

The presence of drug traffickers in protected areas prior to the creation of CONANP had an impact that was not considered when an area was designated as protected. Some areas had already been used for decades to produce marijuana and poppy or as drug trafficking territories, as well as for the involvement of rural and indigenous communities in drug production activities.

> The truth is that these people [drug traffickers] have always been in the area, but they used to be very respectful of the work we did (03-Director of protected area, Sinaloa Cartel territory).

> When we were commissioned to the protected area, we began monitoring birds on the islands and islets in the protected area, and we often saw boats with armed men passing by. They carried drugs by boat to the piers, unloaded the drugs on the beach, and loaded the drugs onto trucks [...]. They already had their transport routes in place long before we arrived [...] This never showed up in the news, but we saw that often (06-Director of protected area, Tijuana Cartel territory).

In addition, environmental inspectors and directors were discreet about information on the location of drug production sites and settlements, as well as trafficking routes within and near protected areas, not out of loyalty to drug traffickers, but to avoid reprisals against the institution, themselves, and the communities. During the period of the pax narcotica, potential complaints were meaningless if the entire police and judicial apparatus was economically dependent on drug trafficking through "economic incentives" and bribes.

According to statements by environmental inspectors and directors, relations with drug producers were always neutral, seeking to avoid any conflict related to illegal activities and prioritising environmental conservation. In several protected areas, especially those far from urban centres, the extent of the fight against organised crime at the national level was not always known, nor was the political pressure of bilateral agreements between Mexico and the United States.

> When we were working with communities on training programmes, during meetings the residents themselves warned us not to enter the protected area because there were anti-drug operations going on [...] they knew and we didn't (04-Director of protected area, Sinaloa Cartel territory).
>
> Sometimes in the morning we would get ready to go up the mountain, and on the way, we would be stopped by an army checkpoint and prevented from passing because there was an operation to destroy marijuana plantations [...] we had to come back later, even up to a week later (07-environmental inspector, territory of the Los Valencia Cartel).
>
> We hardly ever heard about large-scale operations like the ones we experienced later [...] At that time, the whole team was involved in conservation activities, some working with communities and others in the field monitoring wildlife. We welcomed students from the university and there was no danger (08-Director of protected area, Sinaloa Cartel territory).

This situation is characteristic of the pax narcotica period. Drug enforcement operations during this period were selective and only motivated by pressure from cooperation agreements with the United States. Furthermore, limited information was available locally regarding the political and security situation at the national level. Consequently, environmental conservation activities within protected areas continued to strengthen and collaborative links were established with communities within and near protected areas.

Working with Communities Inside and Nearby Protected Areas

CONANP personnel generally conducted environmental conservation activities within protected areas with the support of members of rural and indigenous communities, which fostered a close relationship that facilitated the implementation of environmental conservation activities. These activities included not only wildlife monitoring, but also sustainable development activities such as strategies for sustainable farming and livestock breeding, as well as the use of forest resources such as land, water, timber, and wildlife, and the creation of brigades to respond efficiently to forest fires.

> In those years, we could work without any problems, as long as we didn't get involved in their business. We did everything we could for environmental conservation and to work with the communities. Back then, we all knew who was involved in those activities [drug trafficking], and generally everyone in the communities already knew who we were and what we were doing there, so they didn't bother us. (02-Environmental inspector, Sinaloa Cartel territory).
>
> There was always a very good relationship with the communities, but many people from the community also worked in the fields, many were "gomeros" (poppy gum producers), and we tried to alternate conservation activities with their agricultural activities involving drug production. For example, they always participate in firefighting brigades and courses on organic farming, water efficiency, etc. (04-Director of protected area, Sinaloa Cartel territory).

In rural and indigenous communities, the production of drugs such as marijuana and poppy was the main source of economic income. In the Golden Triangle region, the economic windfall during the harvest allowed families and entire villages to earn enough money to get through the winter season, characterised by snowfall and freezing temperatures.

> I remember when I asked one of the young people in the community why he worked as a "chutamero" [poppy gum producer and drug trafficking] and he said: If I planted corn and beans that gave me the same profits as poppies, then I would plant corn and beans. What earns me money to support my family is the poppy (02-Environmental inspector, Sinaloa Cartel territory).

In addition, the leading drug producers performed functions typical of government security forces, administering justice (through beatings, threats, or death) to those who committed violent acts, theft, or caused harm to members of the community, but they also supported improvements to housing, restoration of roads, and the promotion of grocery stores, hardware stores (ironmongery), etc.

> In one way or another, at that time people lived well on the money they earned from growing drugs, and hardly anyone migrated to the city as happened later [...] the drug traffickers always paid well, and since only one cartel ruled the area, everything was peaceful. Except when there were anti-drug operations, life on the ranches would come to a standstill for a while, and people wouldn't even leave their homes. What I can tell you is that the people [the drug traffickers] helped them a lot. They brought electricity, televisions, water pipes and all those things that the government had never done to the communities (02-Environmental inspector, Sinaloa Cartel territory).

The presence of drug traffickers within and on the peripheries of protected areas represented an economic windfall that was rarely matched by other jobs in the regions. In addition, social discontent over the government's neglect of social issues widened the gap between the government and its population, so that people often felt empathy towards drug trafficking leaders because of their "good deeds" for social development. This was evident in the "protection" of producers and leaders when anti-drug operations were implemented near protected areas.

> When the anti-drug operations arrived with army and police, people always said they didn't know who had poppy fields. But it was the drug traffickers who bought from their grocery stores, who paid them to grow poppies and marijuana [...] How do you think they were going to turn them in? The drug traffickers had brought electricity to the village, they had brought them water and defended them from any danger [...] even the municipal police who accompanied the army received money [...] the whole community lived well thanks to them [the drug traffickers] (08-Director of protected area, Sinaloa Cartel territory).

Environmental conservation activities were conducted in a context where the cultivation and harvesting of drugs was part of community life in indigenous and rural communities, but also part of activities within protected natural areas. In this context, when CONANP personnel had to enter protected areas, particularly territories that were close to poppy or marijuana fields or located on drug trafficking routes, inspectors would notify drug trafficking leaders about the activities they were going to implement in those territories. This was not for the purpose of informing them, but rather to avoid any incidents that might lead to them being confused with other government entities in the field.

> Over time, we learned to work alongside drug traffickers in areas where there were drug plantations [...] but we tried to involve the communities as much as possible in productive practices that would not cause environmental harm and would generate income for them. But I think what helped us most was communicating with local drug traffickers' leaders to let them know about the activities we would be conducting within the protected area [...] not because we had to notify them, but to prevent them from mistaking us for intruders and attacking us while we were working in the area. [...] this also helped us gradually gain the trust and participation of the community members (01-Environmental inspector, Sinaloa Cartel territory).
>
> When we went bird watching, we would notify someone from the community so they could warn the cartel leaders [...] we didn't want any surprises or scares so far from the shore or help [...] we couldn't report anything, we just had to adapt and do our job. What I can tell you is that it wasn't dangerous for us; in those days we worked in peace, but you have no idea how things got later on (10-environmental inspector, Tijuana Cartel territory).

However, alongside environmental conservation activities within protected areas, CONANP's framework was consolidated during this period, strengthening the National System of Protected Natural Areas (SINANP), decreeing new protected areas, generating knowledge through scientific research, working with indigenous and rural communities on local development, and strengthening environmental legislation. These activities were implemented in parallel with government attempts (both domestic and foreign) to combat drug trafficking and the struggles between cartels to dominate more territory and impose over other cartels.

Environmental Conservation Amidst Cartel Clashes

Considering that several protected areas had historically shared territories with drug trafficking and communities involved in drug production, mainly in the north of the country, and specifically in the Golden Triangle, CONANP personnel adapted to this context and conducted environmental conservation and social development activities with the communities. However, in other parts of the country where other criminal groups exercised their dominance, this situation did not last long.

When clashes between cartels intensified as they sought to consolidate their power and acquire more territory, protected areas served as refuges for sicarios. They went from being merely territories for planting, harvesting, and transporting drugs to also becoming refuges and tactical training camps.

> When "Los Chapos" [Sinaloa Cartel] wanted to expand into this part of the state, this region, including the protected area, became a war zone. They wanted to expand to Coahuila to reach the Tamaulipas border, but they were stopped by the Gulf Cartel. There were times when we saw convoys full of armed people passing through the protected area, or remains of campfires in the mountains where the sicarios had slept, tyre tracks from large vehicles and gunshots during the night [...] at that time, we were afraid to do this [environmental conservation]; we were unarmed and defenceless in a territory disputed by two cartels (09-Director of protected area, Gulf Cartel territory).
>
> The first time we saw a drug traffickers' camp inside the protected area, we knew it was a point of no return. They had discovered the benefits of nature, not only to hide wildlife from humans, but also to hide humans from other humans [...] Thank God that camp was empty, otherwise I don't even want to think about what would have happened. They would have killed me and raped and killed my female colleague [...] We had already been warned that the people from Los Zetas were violent [...] We never returned to that area of the protected area (11-environmental inspector, territory of the Zetas Cartel).

The presence of criminal groups in protected areas during this period differs from that of the 1980s, 1990s, and early 2000s. The political and security situation at the government level and the emergence and expansion of other emerging cartels have made protected areas dangerous for CONANP personnel and, consequently, for environmental conservation.

The main risks to environmental conservation in protected areas for CONANP personnel were threats, injuries, homicides, and theft. However, although there are no official records of attacks on CONANP personnel during this time, the situation worsened in the following decades. Even so, the experiences of inspectors and directors reveal the visible and latent risks to environmental conservation posed by the presence of drug trafficking in protected areas during the first 6 years of the twenty-first century in Mexico.

As a clarification, CONANP is a government administrative institution (CONANP, 2018), which means that its personnel cannot carry firearms, even for self-defence, as the Ley Federal de Armas de Fuego y Explosivos (2025) (Federal Firearms and Explosives Law) establishes that the use of firearms must be prioritised for public servants involved in public and private security.

Furthermore, CONANP personnel are not trained for armed confrontations, nor were they trained in contingency plans for violent events such as armed confrontations, self-defence, or any other public safety training. This situation represented one of the main limitations to implementing the activities required by their jobs in the presence of violent drug traffickers in protected areas.

However, the discussion should not focus on whether CONANP personnel should carry firearms for self-defence, but rather on the context of the presence of drug traffickers in protected areas designated for environmental conservation and on the imminent struggle between cartels to increase their control over more territory and government pressure on drug production and trafficking through Mexican territory.

Impact of Drug Trafficking on Protected Areas, Some Distinctions

Considering the experiences of CONANP directors and environmental inspectors during this period, each cartel can be differentiated in terms of its influence on environmental conservation activities in protected areas. In the early years of Vicente Fox's 6-year term, the Sinaloa Cartel sought to consolidate and (re)position itself as the leader of drug trafficking in Mexico (after El Chapo Guzmán's escape from prison), but it was confronted with the emergence of other cartels such as the Gulf Cartel and the consolidation of the Zetas Cartel and the Tijuana Cartel.

However, according to the experiences of CONANP directors and inspectors, the situation within protected areas was barely affected, and environmental conservation activities and work with rural and indigenous communities within and near protected areas continued. Although communities continued to plant and harvest poppies and marijuana, in the second half of the 6-year term, there was evidence of an attempt by the Sinaloa Cartel to expand its territory eastward into the northern states, which led to the protected areas located on the cartels' territorial boundaries becoming disputed territories, leaving CONANP personnel in the middle of armed confrontations.

In the case of the Tijuana Cartel, the working situation for CONANP personnel was different. Threats to interfere with drug trafficking activities or in communities determined how they worked. The inspectors adapted and established contact with community members, whom they informed about the conservation activities they would implement, and in turn with the cartel leaders, which allowed CONANP personnel to partially implement their work agenda and the management programmes for each protected area. This situation corroborates the violent profile of the Tijuana Cartel, as documented by historians of drug trafficking in Mexico (Astorga, 2005; Grillo, 2011; Valdés, 2013) and which must also be understood in the historical context in which the Sinaloa Cartel and its allies, such as the Juárez Cartel and the national government, had among their main objectives the elimination of the Tijuana Cartel.

The consolidation of the Gulf Cartel in drug trafficking and its armed faction, the emerging independent cartel Los Zetas, represented a fundamental factor not only for public security but also for environmental conservation in protected areas. During this period, there were relatively few experiences of inspectors and directors of protected areas interacting with this cartel for two reasons: (1) its recent creation and incorporation into drug trafficking, and (2) its expansion into new territories was just starting.

Even so, the experiences recounted by CONANP personnel document that the Gulf Cartel, in its attempt to consolidate and strengthen its influence in northeastern Mexico, made the protected areas on the borders of its territories dangerous for CONANP personnel due to the camps that were set up inside and on the outskirts of the protected areas, as well as the presence of convoys of armed sicarios who monitored the areas in an attempt to halt the expansion of the Sinaloa Cartel.

Furthermore, Los Zetas, as the armed faction of the Gulf Cartel, spearheaded the expansion of the Gulf Cartel's territory, eliminating smaller cartels and drug traffickers in the process and causing a wave of violence wherever they went and established themselves. This included protected areas where environmental conservation activities were already being implemented or new protected areas were being decreed, with sicarios settling in rural communities near the protected areas.

During this time, CONANP personnel were aware of the violence that characterised this criminal group, as well as its diversification of criminal activities beyond drug trafficking (mainly extortion, kidnapping, and murders), which, according to Valdés (2013), was due to the Gulf Cartel's inability to pay all its members' salaries, and they began to generate their own income through other illegal activities that were not drug trafficking. This posed a danger to the implementation of protected area management programmes in the regions invaded by Los Zetas.

The resilience of CONANP personnel meant that, even with the security situation in the early 2000s, management programmes could be implemented in protected areas, and communities could be involved in environmental conservation and sustainable regional development activities. Inspectors and directors of protected areas implemented these programmes in the context of relative peace, but which was already at its sustainable limit.

In conclusion, the environmental conservation scenario in protected areas of Mexico during the first 6 years of the 2000s was characterised by the emergence of CONANP as a new government environmental institution that joined a complex scenario in which government institutions at all three levels of government (municipal, state, and federal) had to align themselves with the objectives set by the federal government, which in turn was under pressure from the United States to fight drug trafficking (Fig. 7.3).

Conservation activities in natural areas alternated with periods of planting, harvesting, and drug trafficking, in which communities also participated and which represented their main sources of economic income. Furthermore, with the emergence of new cartels during this time, CONANP personnel's strategies for adapting to this scenario could not be standardised at the national level; strategies that worked with one criminal group did not work with others.

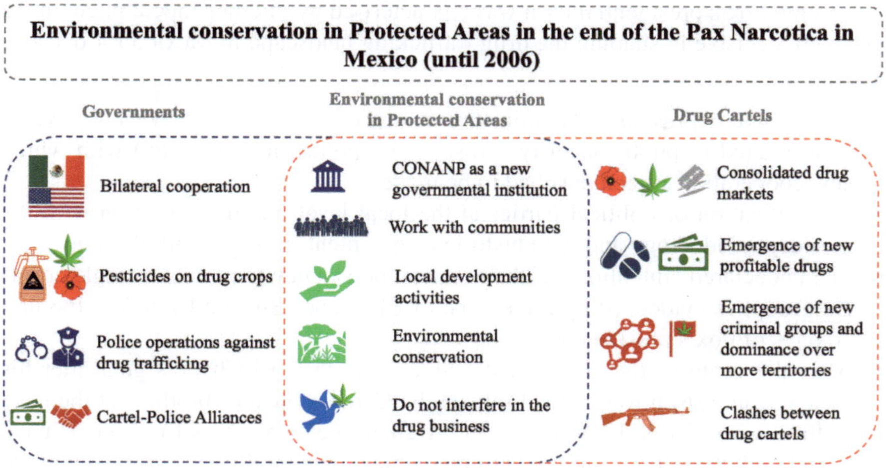

Fig. 7.3 Scenario of environmental conservation in protected areas in the context of the end of the pax narcotica in Mexico and the prelude to the "war on drugs". *Source*: The authors

However, the experiences lived and learned by CONANP personnel during this period were only the prelude to environmental conservation and the implementation of management programmes in protected areas in a war scenario, which was experienced during the following 6-year government term. This demonstrates the resilience of CONANP personnel, not only in implementing conservation activities (wildlife monitoring, forest restoration, fighting forest fires, etc.), but also in involving rural and indigenous communities in environmental conservation and promoting activities that encourage sustainable development at the local level.

The Prelude to the "War on Drugs" in Mexico, the End of the Pax Narcotica

The "new" and old cartels continued to operate in drug trafficking during the latter half of the 1990s and the first 6 years of the 2000s. At the end of Ernesto Zedillo's presidential term (1994–2000) and the beginning of Vicente Fox Quesada's presidential term (2000–2006), the drug trafficking scene was influenced by a change in Mexico's political position.

After 70 years of political dominance by the Partido Revolucionario Institucional (PRI), the era of PRI presidentialism came to an end with the arrival of Vicente Fox as president. The dominant political party was the Partido Acción Nacional (PAN), with a right-wing, conservative political doctrine and a clear idea of tackling drug trafficking in Mexico, or at least, although not explicitly, to try to consolidate a single dominant group that could be controlled or have strong alliances with the government.

Vicente Fox's presidential term was characterised by interdependent phenomena that were decisive in shaping the drug trafficking landscape in Mexico for decades to come:

- The change in presidential administration and changes at other levels of government related to public security (prosecutors, police, military, etc.) with which new cooperation agreements had to be made
- Diversification of political parties at the local level that did not align with the presidential ideology and with historical agreements with organised crime
- The undeclared "intention" of the Mexican presidency to protect a single criminal group, as evidenced by the escape of El Chapo (Sinaloa Cartel) in the first 20 days of Fox's presidency
- After the death of the "Lord of the Skies" in 1997, El Chapo's goal after his escape from prison was a) to eliminate the Carrillo Fuentes brothers at the head of the Juárez Cartel, b) eliminate the Arellano Félix brothers from the Tijuana Cartel with the support of a government crackdown on the Tijuana Cartel at the beginning of the 6-year term, and c) capture Osiel Cárdenas, leader of the Gulf Cartel, in 2003. These three events favoured the expansion of the Sinaloa Cartel during Fox's 6-year term.
- Violent attacks, mainly homicides and extortion, against civilians (business owners, construction workers, grocery stores, workshops, etc.) by organised crime groups seeking to control more territory and impose dominance over other cartels, such as the Los Zetas cartel.
- The consolidation of new drugs such as amphetamines and methamphetamines in the global drug market, with western regions of the country being the main producers of these drugs, mainly in the states of Jalisco, Michoacán, and Guerrero, but also in the central part of the country, such as Mexico City and the State of Mexico, through the creation of drug laboratories in urban areas, which in subsequent decades would migrate to rural and protected areas

This scenario laid the groundwork for a policy of direct combat against drug trafficking in Mexico during the following 6-year term. However, it should be noted that by the end of Vicente Fox's term in 2006, there were 160 protected areas, meaning that 10 new protected areas were created in Mexico during this 6-year period. As a result, CONANP personnel in the new protected areas faced the context of the Mexican War on Drugs without the prior experience of other protected areas.

Conclusions

In the Mexican drug trafficking scenario during this period, in addition to police officers, commanders, military personnel, prosecutors, and public ministers (all of whom are involved in public safety), a new government figure has emerged who interacts directly with these criminal groups: public servants dedicated to environmental conservation, such as environmental directors and inspectors of protected areas (all of whom are involved in environmental conservation).

The interaction between a public security issue (drug trafficking) and institutional efforts to conserve the environment adds another layer of complexity to the study of drug trafficking in Mexico. This raises the question of whether drug trafficking in Mexico's protected areas should be considered a public security or an environmental issue (Carpio-Domínguez, 2021).

A first approach to answering this question is to consider that protected areas are public spaces and, therefore, public safety must be guaranteed, understood as a quality of public (and private) spaces characterised by the absence of threats that undermine or threaten human rights and in which the necessary conditions exist for the development of personal existence (García, 2002). Building this quality involves the participation of multiple resources from society and the State (Centro de Estudios Sociales y de Opinión Pública, 2006).

Protected areas are territories that contribute to guaranteeing a Mexican constitutional right, the right to a healthy environment (Article 4 of the Constitution of the United Mexican States [Constitución de los Estados Unidos Mexicanos] 2025). Therefore, the presence of criminal groups threatens environmental conservation efforts and, consequently, biodiversity and ecosystems. Furthermore, criminal activities such as drug trafficking and other emerging activities such as extortion, threats, and homicides limit the development capabilities of human communities in these areas.

For Green Criminology, the line separating criminal phenomena of public security from environmental crimes and harms is not clearly defined, given that environmental harms cut across other forms of crime (Carpio-Domínguez, 2023), as evidenced by drug trafficking activities in protected areas of Mexico.

Therefore, environmental conservation actions in Mexico during this period were conducted in the context of relative pax narcotica, but they also preceded one of the most violent periods in the country's history: the Mexican War on Drugs (2007–2012), which influenced environmental conservation efforts amidst shootings, killings, curfews, drug crops, and death.

References

Astorga, L. (2005). *El siglo de las drogas. El narcotráfico, del Porfiriato al nuevo milenio*. Plaza y Janés.

Bernard, R. (2006). Nomprobability sampling and choosing informants. In R. Bernard (Ed.), *Research methods in anthropology. Qualitative and quantitative approaches* (pp. 186–209). AltaMira Press. https://www.zambiancu.org/1zRead/Bernard-ResearchMethodsInAnthropology%20%5Bxoer%5D.pdf

Carpio-Domínguez, J. (2021). Crimen Organizado (narcotráfico) y conservación ambiental: el tema pendiente de la seguridad pública en México. *Revista CS, 33*, 237–274. https://doi.org/10.18046/recs.i33.4076

Carpio-Domínguez, J. (2023). *Criminología verde para México: Desarrollo de una perspectiva verde en la herencia criminológica mexicana*. Universidad Autónoma de Tamaulipas-Tirant Lo Blanch. https://doi.org/10.29059/LUAT.317

Centro de Estudios Sociales y de Opinión Pública. (2006). *Seguridad Pública*. Gobierno de México. http://archivos.diputados.gob.mx/Centros_Estudio/Cesop/Comisiones/dtseguridad%20publica1.htm#[Citar%20como

Comisión Nacional de Áreas Naturales Protegidas (CONANP). (2015). *Áreas Naturales Protegidas*. Regiones. Gobierno de México. https://www.gob.mx/conanp/acciones-y-programas/areas-naturales-protegidas-51333

Comisión Nacional de Áreas Naturales Protegidas (CONANP). (2018). *Plan de Acción para el Financiamiento Estratégico 2016–2018*. Gobierno de México. https://www.conanp.gob.mx/Turismo/PAFESegundaEdicion2017.pdf

Comisión Nacional de Áreas Naturales Protegidas (CONANP). (2022). *Sistema Nacional de Áreas Protegidas (SINAP)*. Gobierno de México. SEMARNAT. https://www.gob.mx/conanp/acciones-y-programas/sistema-nacional-de-areas-protegidas-sinap

Comisión Nacional de Áreas Naturales Protegidas (CONANP). (2025a). *Programas de Manejo de las Áreas Naturales Protegidas de México*. Gobierno de México. https://www.gob.mx/conanp/acciones-y-programas/programas-de-manejo

Comisión Nacional de Áreas Naturales Protegidas (CONANP). (2025b). *Sistema de Información*. Monitoreo y Evaluación para la Conservación (SIMEC). Gobierno de México. https://simec.conanp.gob.mx/

Constitución de los Estados Unidos Mexicanos. (2025). *Última reforma publicada en el Diario Oficial de la Federación el 15 de abril de 2025*. Cámara de Diputados del H. Congreso de la Unión. Gobierno de México. https://www.diputados.gob.mx/LeyesBiblio/pdf/CPEUM.pdf

García, S. (2002). En torno a la seguridad pública. Desarrollo penal y evolución del delito. In J. Peñaloza & M. Garza (Eds.), *Los desafíos de la seguridad pública en México* (pp. 81–98). Universidad Iberoamericana. https://archivos.juridicas.unam.mx/www/bjv/libros/1/419/10.pdf

González-Ocampo, H., Cortés-Calva, P., Íñiguez-Dávalos, L., & Ortega-Rubio, A. (2014). Las áreas naturales protegidas de México. *Investigación y Ciencia, 22*(60), 7–15. https://www.redalyc.org/pdf/674/67431160002.pdf

Grillo, I. (2011). *El Narco: Inside Mexico's criminal insurgence*. Bloomsbury Press.

Ley Federal de Armas de Fuego y Explosivos. (2025). *Última reforma publicada en el Diario Oficial de la Federación el 29 de mayo de 2025*. Cámara de Diputados del H. Congreso de la Unión. Gobierno de México. https://www.diputados.gob.mx/LeyesBiblio/ref/lfafe.htm

Valdés, G. (2013). *Historia del narcotráfico en México*. Editorial Aguilar.

Chapter 8
Drug Trafficking in Protected Areas: During and After the Mexican War on Drugs (2007–2012) and the Post-War Period (2012–2018)

Introduction

With a tense drug trafficking scenario at the national level, the Mexican War on Drugs emerges. With the end of Vicente Fox Quesada's presidency, President Felipe de Jesús Calderón Hinojosa wins the elections for the presidential term from December 2006 to November 2012. Although in literature and the media this period is considered one of the most violent in Mexico's history in terms of public security, under the discourse of the fragmentation of criminal groups resulting from the government's fight against drug cartels (Chabat, 2014).

However, the fragmentation of criminal organisations began in the late 1980s, and according to historical records (Astorga, 2005; Grillo, 2011; Valdés, 2013), this fragmentation had several causes, the first caused by the arrests of Miguel Ángel Félix Gallardo, Ernesto Fonseca "Don Neto", and Rafael Caro Quintero, which led to the distribution of "plazas" among the leaders and some public servants of the Federal Judicial Police.

Another cause of fragmentation was the disappearance of economic barriers, as there was no single criminal group wielding dominance over the drug trade. This led to the emergence of new criminal groups such as the Gulf Cartel and La Familia, also known as La Familia Michoacana, and some other groups that did not last long, such as Los Valencia in Michoacán and Los Amezcua Contreras in Colima.

However, after 80 years of uninterrupted drug trafficking in Mexico, criminal activities related to drug trafficking were operating nationwide in all states of the country, but also in other parts of Latin America, Europe, the United States, Canada, and Eastern countries such as China and Japan, among others.

The start of President Felipe de Jesús Calderón Hinojosa's six-year term and the public security policies implemented during the previous term were the preamble to

Supplementary Information The online version contains supplementary material available at https://doi.org/10.1007/978-3-032-18218-0_8.

J. L. Carpio-Domínguez et al., *Drug Trafficking and Protected Areas in Mexico*, https://doi.org/10.1007/978-3-032-18218-0_8

an imminent war. Conflicts over territory, the expansion of criminal groups, and violent forms of imposing dominance over other cartels and their territories, along with the implementation of activities other than drug trafficking by criminal groups such as Los Zetas, including murders, kidnappings, extortion, and deprivation of liberty, among others, undermined the population's sense of security.

It was no longer just a violation of the rule of law at the national level or a drug production practice with no major impact on homeland. At the beginning of this presidential term, attacks against the civilian population began to be reported in the media and attracted even more attention from national and international governments. In addition, the scenario of bilateral collaboration between Mexico and the United States, plus the political orientation of the new political party and its intention to confront the cartels (whether genuine or forced by bilateral agreements with the United States), created a scenario conducive to war.

During this time, environmental conservation efforts by the National Commission for Protected Natural Areas (CONANP) personnel, who already had 6 years of experience in developing environmental conservation activities in protected areas occupied by drug traffickers, were also consolidated. However, the dynamics for implementing environmental conservation activities changed, not only in terms of the activities that could be implemented within protected areas, but also in terms of collaborative relationships with rural and indigenous communities in the context of armed conflict, excessive violence, convoys of sicarios, drug cultivation fields, and drug traffickers' camps.

This chapter explores the experiences of environmental inspectors and directors of protected areas in Mexico, considering a diachronic analysis during the Mexican War on Drugs in the period 2007–2012, which allows us to contrast the situation before the war and the post-war scenario. In addition, documentary analysis is used to provide a historical context for the events that shaped the homeland security situation, which is then contrasted with the experiences of CONANP personnel to highlight the working conditions of environmental conservation in protected areas during the Mexican War on Drugs.

However, it is necessary to clarify that this chapter is not intended to be a historiographical analysis of drug trafficking in Mexico during this period, but rather to highlight historical events in order to document the experiences of CONANP personnel in the context of the Mexican War on Drugs.

Public Safety Scenario During 2007–2012

Considering that criminal groups involved in drug trafficking operated in all states of the country, it is difficult to identify the specific municipalities in which they operated; however, to a greater or lesser extent, local leaders, public servants, and civilians in most municipalities in each state had some type of relationship or connection with some criminal group.

According to Valdés (2013), the Sinaloa Cartel (also known as the Pacific Cartel) exercised control over all the coastal states of the Pacific Ocean, from Oaxaca to Sonora, with the exception of the state of Michoacán and also controlled states such as Durango, Baja California Sur, the State of Mexico, Puebla, Morelos, Jalisco, and Mexico City. The Tijuana Cartel, commanded by the Arellano Félix family, exercised iron-fisted control over the state of Baja California (not to mention their associations with small-time leaders and traffickers in other states along the Pacific coast).

During this time, the Gulf-Zetas Cartel also emerged, exerting control from the state of Tamaulipas in the north of the country to Chiapas in the south, including Nuevo León, Coahuila, San Luis Potosí, Zacatecas, Aguascalientes, Hidalgo, Veracruz, Tabasco, Campeche, and Quintana Roo. The Michoacán Family fought Los Zetas for control of the state of Michoacán (by this time, the Valencia Cartel had been dismantled by Los Zetas and government operations). In addition, La Familia began to exert control over the states of Guanajuato, Guerrero, Querétaro, and Mexico State. Meanwhile, the Juárez Cartel ruled over states such as Chihuahua and Quintana Roo (Valdés, 2013).

The territorial distribution of these criminal groups shaped the drug trafficking scenario at the beginning of Calderón's administration, with two clear objectives: a) to become the dominant criminal group and b) to remain in a highly profitable market regardless of the human and social cost.

To understand the importance of the geographical distribution of criminal groups involved in drug trafficking, it is necessary to understand that during this period, drug cartels continued to operate as large national and international criminal enterprises due to two main factors: a) Diversification within the drug market: marijuana, cocaine, heroin, and methamphetamine; and b) the cartels had the infrastructure to produce, transport, and smuggle these drugs into the United States.

The old and new drug cartels organised drug production with the participation of thousands of farmers, creating and operating drug laboratories for heroin and methamphetamines with a system for supplying imported raw materials (mainly from China and other East Asian countries), as well as collection, storage, and transport systems (maritime, land, and air).

This, combined with the large size of Mexico's territory, characterised by an orography that facilitates these activities but also conceals them from national and foreign government actions. Furthermore, it should be noted that the incursion of cocaine into the drug market gave Mexican cartels an international influence that few could have foreseen, and which remains to this day.

Furthermore, with the transition of presidential administration and since the collapse of the Sinaloa Organisation and its relationship with Mexico's General Security Agency, relations between drug cartels and their leaders and government security organisations have resulted in a set of relationships at the local (municipal) and state levels that are very difficult to document. However, there is no doubt that during this period, drug cartels in Mexico required government protection, mainly from municipal and state police, public prosecutors, mayors and municipal presidents, governors, public security secretaries, judges, prison directors, commanders, and state and federal investigative police, but also from the military in many regions of Mexico (Astorga, 2005; Grillo, 2011; Valdés, 2013).

In parallel with government protection, the cartels began to form subgroups within their organisations to carry out specific activities, such as informants, also known as "halcones" (hawks), who were positioned as taxi drivers, petrol station attendants, newspaper sellers, drug dealers, or civilians directly on the street. These "halcones" reported on police operations in the street and the presence of members of other cartels in their territories or acted as a link to civil society to request "support" if there were thieves, swindlers, or extortionists, who were killed, beaten, and/or threatened by cartel sicarios as a form (albeit distorted) of citizen protection.

Armed Commandos Within Drug Cartels

In addition to the network of spies and informants that drug trafficking groups had in Mexico, the emergence of subgroups specialising in armed conflict marked a new phase in Mexican drug trafficking during this period. Groups such as Los Zetas represented an escalation not only in drug trafficking tactics, but also in the violence used to assert their dominance in the drug market and in the territories.

Osiel Cárdenas Guillén, leader of the Gulf Cartel, (unintentionally) provided this group with all the tactical and logistical tools they needed to break away from the Gulf Cartel itself. Once Cárdenas Guillén was imprisoned, the new leaders of the cartel were unable to control the armed group, Los Zetas. Thanks to this group, the cartel had expanded its territory from the northeast to the south of Mexico, managing to confront other more experienced and long-established cartels using violence and narco-terrorism as their main strategies.

Los Zetas were the first to diversify their criminal activities beyond drug trafficking, venturing into crimes such as extortion (fees for having shops and other businesses named "cobro de piso"), vehicle theft, kidnapping, human trafficking (migrants), and sicariato (contract killings), among others, which were characterised by the military and organisational capacity (through cells) of this cartel. During the period of expansion of this cartel, others such as Los Valencia were wiped out, and they exercised control over the territories with an iron fist and extreme violence.

To counter these operational methods of Los Zetas, the other cartels also formed more specialised armed subgroups. The Sinaloa Cartel founded the group known as "Los Pelones" in the service of El Chapo, the Tijuana Cartel founded "FEDA" (Arturo's Special Forces), and the Juárez Cartel founded "La Línea" as a defence group, while within Los Zetas, "La Familia" emerged. In addition, the use of gang members such as "Los Mara Salvatrucha" as frontline soldiers within the forces of Mexican drug trafficking began to be documented.

At the end of 2006, when Felipe de Jesús Calderón Hinojosa took office as president, he found a country in which five major drug cartels exercised control over almost all social, political, and security spheres. The cartels were characterised by their use of violence as a combat strategy, rather than agreements and pacts, with the aim of exercising their dominance over as much territory as possible, as well as occupying emerging drug markets or any business that offered large financial gains.

Furthermore, according to Valdés (2013), in the United States, President George Bush did not renew the ban on the sale of automatic and assault weapons imposed 10 years earlier by President Clinton. This had a significant impact on Mexico because, although drug traffickers could purchase any firearm on the black market before this policy, they could now obtain them at more affordable prices and without legal resources in the more than 11,000 gun shops located in the United States less than 100 miles from the border. Considering the security situation in Mexican territory, this only made it easier for drug traffickers to access more firearms.

Calderón's Mexican War on Drugs

In the social imaginary, Felipe Calderón's public security policy against drugs is associated with (directly or indirectly) causing the war and fragmentation of the cartels. However, since the 1980s, clashes over territory in Tijuana, Nuevo Laredo, Ciudad Juárez, and Michoacán, as well as government operations to imprison leaders, led to three "wars" being underway by the end of 2006 (Valdés, 2013, p. 368): a) the Tijuana Cartel against the Sinaloa Cartel, b) the Gulf Cartel against the Sinaloa Cartel, and c) the Los Zetas Cartel against the La Familia Cartel. Added to this is the specialisation of each cartel in weapons and operations.

In this scenario, the diversification of activities by the cartels represented a turning point in relation to the government's stance towards them. The Zetas began looking for new sources of income alternative to drug trafficking, which they had learned from the Gulf Cartel. In addition to extortion ("cobro de piso"), vehicle theft, kidnapping, human trafficking (migrants), and sicariato, they learned how to form a network for stealing hydrocarbons to supply their own vehicles, but also for sale at petrol stations through a network that included public servants and the private sector.

The illegal extraction of hydrocarbons is known as "huachicoleo" and is perpetrated through clandestine taps on the State-owned Petróleos Mexicanos (PEMEX) pipeline network. This phenomenon, which emerged in recent years, would decades later become an illegal trafficking market almost as significant and profitable as drugs, mainly in the states on the Gulf of Mexico coast.

However, the Tijuana and La Familia cartels also diversified their activities, filling their coffers and strengthening their position with money from society through extortions (cobro de piso), vehicle theft, kidnapping, and sicariato. The impact of these activities on local economies in many municipalities across the country created an impression (social perception) that had never been seen before and, as a result, widespread (social) discontent.

This led to the closure of small and medium-sized businesses (grocery stores, auto repair shops, tyre shops, petrol stations, tortilla shops, beauty salons, etc.) that were unable to sustain themselves and pay fees to criminal groups, undermining the diversity of economic sectors at the national level. This situation was exacerbated by the links between criminal groups and public servants who offered them protection, often through bribes and agreements, but during this time a new form of relationship was also introduced: threats. As a result, public servants protected members of organised crime due to threats of reprisals.

Another important change in the drug trafficking scenario in Mexico during this time was that it was no longer possible to generalise all groups under the term "drug traffickers" or "narcos", because not all of them were involved in drug trafficking. Since 1996, criminal association has been classified as a crime in Mexico's Federal Law against Organised Crime, defined as "when three or more persons organise themselves to perpetrate, on a permanent or repeated basis, acts which, individually or in combination with others, have as their purpose or result the commission of crimes ranging from terrorism to crimes against the environment" (Ley Federal contra la Delincuencia Organizada, 2021). This would be one of the main legal frameworks that would allow justice to be enforced in relation to drug trafficking and other cartel activities in the following decades.

Four years later, at the Palermo Convention, the United Nations (2000a) defined organised crime as:

- A structured group of three or more persons
- The group exists for a certain period of time
- It acts in concert with the aim of committing one or more serious crimes
- With the aim to obtain, directly or indirectly, an economic or other material benefit

Mexico's participation in the Palermo Convention (United Nations, 2000b) allowed it to consolidate its stance on drugs with the aim of confronting the cartels that were now not only trafficking drugs but also joining forces to attack society, its property, and personal safety.

Although there was no "official" declaration of war on drug trafficking, the set of public security policies influenced the name given to this presidential term. The start of Felipe Calderón's presidency on 1 December 2006 was marked by "Operativo Conjunto Michoacán" (Joint Operation Michoacán) on 11 December of that same year, just 10 days after he became president. This operation consisted of sending 6500 members of the National Army, the Navy, and the Federal Preventive Police to the state of Michoacán to directly combat drug trafficking groups in the state, mainly Los Zetas and La Familia, who controlled most of the state's territory.

This first massive military deployment in a public security operation is considered the starting point of the Mexican War on Drugs—through the declaration that drug trafficking was a homeland security threat and through the interpretation of Article 89, Section VI, of the Political Constitution of the United Mexican States, which establishes that the Armed Forces (Army, Navy, and Air Force) have three main functions:

- External defence
- Homeland security
- Natural disaster response

This allowed the Armed Forces to intervene in restoring public security throughout the national territory, with drug trafficking being declared a homeland security threat rather than a public health issue. Subsequently, in 2007, the Mérida Initiative was launched with the aim of strengthening cooperation between the United States and Mexico to combat drug cartels through economic support and military equipment from the United States, with a budget of around 400 million dollars to be released as operations against drugs progressed in states such as Baja California, the Golden Triangle, Sinaloa, Guerrero, Tamaulipas, and Michoacán in the next few years (Estrada, 2012).

The 2007–2012 Public Security Sector Programme for this six-year presidential term established that the Government of the Republic aimed to restore the original purpose of the public security function in two dimensions: a) for individuals, who expect protection of their physical integrity, their rights, and their property, and b) for society, safeguarding public liberties, order, and peace (Gobierno de México, 2007).

Considering the institutional limitations in public security, much of the effort focused on consolidating public security institutions, seeking to eliminate corruption, but also to improve the ways in which justice was enforced in the country. These efforts were evident in reforms and institutional strengthening, such as the creation of the Federal Police by unifying several federal police forces into a single institution that was well trained, professionalised, and subject to confidence control procedures for hiring.

In addition, this confidence control procedures in police recruitment were implemented in the police forces of the three levels of government (municipal, state, and federal), in which many police officers failed the tests and were dismissed. However, one of the unexpected results was the lack of replacements for the dismissed police officers, which was evident in a shortage of active police officers at the national level.

Another major advance in public safety implemented during this six-year term was the 2008 Criminal Justice Reform, which introduced a new criminal justice system, migrating from an inquisitorial system to an adversarial and oral system through a gradual transition, with a deadline of June 2016 for all states in the country to implement this new justice system.

This justice system allowed for greater transparency and efficiency in judicial proceedings, and although it is not considered a strategy in the fight against drug trafficking, it was promoted as part of the federal government's strategies to law enforcement (Gutiérrez-López, 2016; Capó-Ramírez, 2021). However, the implementation of this justice system was controversial because it included processes and procedures never seen in the prosecution and administration of justice, resulting in the release of individuals prosecuted for various crimes due to due process violations. Although these errors decreased with the experience of all parties involved in criminal justice, several sectors of society had a negative perception of the government's security strategies (Rodríguez-Ferreira, 2012; Moreno-Hernández, 2025).

Even with all these improvements in the governmental scenario, conflicts and adjustments in the organisation of each criminal group continued to brew within the drug cartels, making it increasingly dangerous to live in areas of conflict between cartels. These conflict zones were generally located in border areas and cities, in large cities such as Monterrey, Guadalajara, and Mexico City, in areas of states such as Michoacán, Tamaulipas, and the Golden Triangle, and in places where the leaders of criminal organisations resided.

Geographical Distribution of Drug Cartels in Protected Areas of Mexico

Conflicts were also present in areas that were not documented or featured in the mainstream media, newspapers, or social networks, such as indigenous territories and protected areas. By 2007, Mexico had designated 164 protected areas throughout the country, where conservation activities were implemented within the protected areas and development activities were conducted in rural and indigenous communities.

The geographical distribution of criminal drug trafficking groups was nationwide (Jenner, 2021; Chabat, 2019). Some states served as receptors for cocaine arriving by sea, both on the Pacific coast and on the Gulf of Mexico, while others continued to produce poppy and marijuana. The border region was the gateway to the United States and the main "plazas" disputed by drug trafficking groups (Hurtado & García-Paz, 2013).

With the emergence of methamphetamines, states such as Jalisco, Mexico City, the State of Mexico, and Michoacán had the highest concentration of drug laboratories producing it. Meanwhile, large areas of the north-central states were used as drug trafficking routes, where the cartels had "safe houses" to store drugs, firearms, and other weapons, as well as their own and stolen vehicles.

Meanwhile, protected areas continued to be designated in areas important for environmental conservation due to their concentration of biodiversity, their ecosystems, and the ecosystem services they provided to human communities. Of the 164 protected areas during this period, 28 were marine areas. Furthermore, it should be emphasised that the presence of drug cartels throughout the national territory facilitated the interaction and incursion of cartels into protected areas (Fig. 8.1).

Although several of the protected areas historically occupied by drug traffickers became coveted spaces for criminal groups during this period, both for drug production and as training camps for sicarios and narco-laboratories, this was "new" for environmental inspectors and directors of protected areas, as they had adapted to working and implementing conservation activities in territories with drug traffickers. However, this new scenario of disputes between criminal groups and public policies against drugs presented new challenges for environmental conservation.

The presence of drug traffickers, criminal groups, and armed conflicts in protected areas is a phenomenon that has been documented not only in Mexico, but also in other Latin American countries such as Peru (Núñez, 1999), Nicaragua (Malamud & García, 2011), Panama (EFE Verde, 2018), Brazil (Olinger, 2013), and Colombia

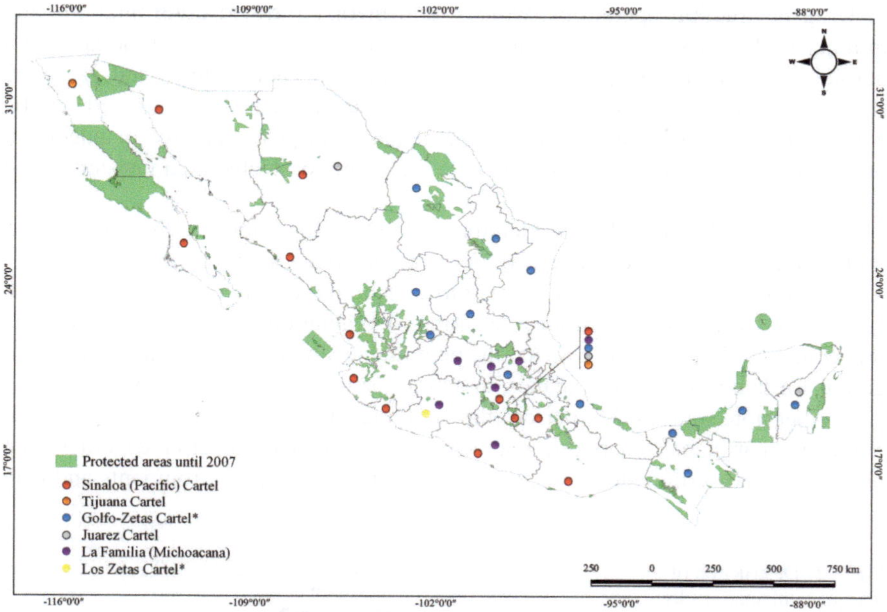

Fig. 8.1 Geographical distribution of protected areas and drug cartels in Mexico until 2007. **Source:** The authors with data from Valdés (2013). **Note:** The circles show the presence of each cartel in each state, not the exact location due to the lack of information on the geographical locations of drug cartels during that time

(Cepeda, 2011; Morales, 2017), where, at some point, criminal groups have dominated large areas of territories dedicated to nature protection.

In countries such as Colombia, armed conflicts were decisive in shaping how territory was understood, occupied, and used (Morales, 2017). Disputes between armed groups (drug cartels and paramilitary groups) and state forces led to the occupation of areas of great natural wealth such as the Amazon, Chocó, the Andes, and Orinoquía, which had a largely negative impact on natural areas to a greater or lesser extent.

Environmental Conservation During the Mexican War on Drugs

The presence of drug traffickers in protected areas not only posed a threat to homeland security, but also to the safety of CONANP personnel. Although they had 6 years of experience working in territories invaded by drug traffickers, the Mexican War on Drugs presented a challenge never seen in environmental conservation in the country.

Drug destruction operations, which often involved armed confrontations and military raids in wilderness and rural areas, were accompanied by broad-spectrum herbicides (pesticides) spraying from light aircraft, helicopters, and by hand (Fig. 8.2). These actions were also implemented in protected areas and nearby zones, which directly impacted the implementation of protected area management programmes, but also the safety of people, including protected area inspectors and directors.

However, it should be highlighted that in Mexico, the impact of these actions on protected areas was not documented. For example, there are no records of the environmental impact caused by the incursions of state forces into protected areas, such as water and air pollution, rubbish, clearing of vegetation to make access possible for security and military forces, and the potential death of animals considered "dangerous for humans", such as snakes, tarantulas, spiders, and wild boars.

Furthermore, during this period, the spraying of broad-spectrum herbicides to eliminate marijuana and poppy plants was not selective but affected other wild plants. The indiscriminate application of herbicides in wild environments and near

Fig. 8.2 Destruction of marijuana plantations by members of the Mexican Attorney General's Office and the Army in Tequila, Jalisco. **Source:** Historical archives of La Jornada (2007)

human settlements was the inflection point for focusing criminological attention not only on American interventionism in Latin American countries (del Olmo, 1987; Goyes & South, 2017), but also on the harm that this practice caused to the environment (biodiversity and ecosystems) and humans.

According to historical records from the daily *La Jornada*, the herbicide used to eradicate poppy and marijuana plants in Mexico was considered "classified" by SEDENA (the Army). However, according to records from the Attorney General's Office, during the period from 2001 to 2006 (Vicente Fox's presidential term), 559 820 litres of Paraquat were purchased, of which 440 381.65 litres were used, leaving a total of 119 440 litres for Felipe Calderón's presidential term (file DGPDSC/UEAI/3380/2006) (La Jornada, 2007).

It is important to remember that Paraquat (Paraquat dichloride) is a highly toxic herbicide, restricted in use and which can only be applied under certification; it is a strong contact herbicide, non-selective and broad spectrum, used to control weeds in various crops as a defoliant. Ingestion of Paraquat can be fatal and contact with the skin or eyes can cause serious effects. In terms of environmental harms, Paraquat is a potential risk to mammals, birds, terrestrial invertebrates, terrestrial plants, and algae (United States Environmental Protection Agency [EPA], 2024).

Paraquat was mainly applied by aerial spraying. During Felipe Calderón's presidential term, the Attorney General's Office provided the military with 119 440 litres of Paraquat, as well as 58 aircraft to apply it in various drug destruction operations. According to La Jornada (2007) in the first report on the activities of Calderón's presidency: "from February to June, the transfer of various air bases was formalised, along with assets associated with eradication, such as furniture, computer equipment, programmes, files, office equipment, herbicides, laboratory equipment, fumigation equipment, aircraft, vehicles and tanker trucks."

The classified status of the information limits the understanding of the dynamics of the application of this herbicide (or possibly others such as glyphosate), as well as the environmental impacts caused, and no academic studies analysing the environmental or human harms have been identified to this moment.

The implementation of these public policies against drugs had an impact that was not quantified in Mexico either during or after the war period. This shows that public safety was prioritised without considering the environmental and human impacts of drug plant eradication strategies. However, although most of these policies were not implemented within protected areas, most of the plantations were located in wild environments far from large urban centres and away from government attention, and near to protected areas.

In protected areas, media attention was lower than in other places for two possible interrelated reasons: 1) protected areas are territories under federal administration, CONANP is the institution that administers the country's federal protected areas, and 2) inspectors and directors belong to a federal government institution, and any complaints or interviews with the media contravened government policies and the government's line of action against drugs.

Nevertheless, the conservation activities implemented by CONANP personnel in this political and public security scenario continued with a resilience rarely seen in government institutions and with a willingness and strength to continue environmental conservation actions in protected areas in the context of the Mexican War on Drugs.

Challenges and Efforts to Sustain Environmental Conservation

The deployment of the state armed forces by presidential order at the national level led them to pay special attention to natural spaces and protected areas due to the presence of poppy and marijuana fields. This vulnerable situation caused inspectors and directors of protected areas to find themselves in situations of conflict and various forms of violence due to armed clashes and anti-drug operations.

The main interventions by security forces documented by inspectors and protected area directors were the destruction of drug crops and drug laboratories, but also in marine areas with protected areas through which drugs were transported by speedboats or boats.

> Sometimes they [drug traffickers] kicked us out of the places where we were monitoring marine wildlife. Armed people [drug traffickers] would arrive in boats and push us out, and we had to stop working and come back another day or the next week (06-Director of protected area, Tijuana Cartel territory).
>
> When things got really dangerous, the official instruction was to avoid exposing ourselves to danger, although they [our superiors] didn't tell us exactly how. What we did was reduce our activities in areas of the protected zone where we knew they were hiding in their camps, and although there were only three of us in the team [a director and two environmental inspectors], we also made sure that two of us always went everywhere together; none of us could go alone (13-Director of protected area, Sinaloa Cartel territory).

The beginning of the violent events was marked by an atmosphere of widespread mistrust among members of criminal groups. Although initially groups such as the Sinaloa Cartel and the Juárez Cartel always worked to support people in rural and indigenous communities, this changed partially during the war, while other cartels never provided such collaboration or support to the communities.

Illegal checkpoints, known in Mexican slang as "retenes", were intensified and set up by criminal groups to monitor who entered or left towns in order to prevent members of other criminal groups from entering and threatening the criminal group's control over its territory.

Academic studies on this situation highlighted the impact of the Mexican War on Drugs in rural areas where there was a network of identities, alliances, and cultures deeply rooted in their territories, which had, by then, been invaded by drug traffickers, the military, and the police in a war they could not fight (see Gledhill, 2003; Maldonado, 2011; Rosen & Zepeda, 2015a, 2015b) and of which they were also victims, along with those who conducted conservation activities in protected areas.

This context of territorial expansion, governmental affronts, political pressure, and mistrust also prevailed in the country's protected areas. The early years of the war were characterised by roadblocks (illegal checkpoints), intimidation, deprivation of liberty, injuries, kidnappings, murders, and armed clashes between cartels and security forces, with civil society caught in the middle.

However, although the impact of the war was more visible in cities, rural and indigenous communities received less media attention regarding the impact of drug trafficking and the war, and the experiences of inspectors and directors of protected areas remained unnoticed (possibly for the reasons outlined previously).

> A fellow inspector who was monitoring wildlife was detained by drug traffickers on his way to the protected area. They took his equipment, including monitoring cameras, herpetological hooks and other tools, but that was all. We all know that in other circumstances it could have been much worse (13-Director of protected area, Sinaloa Cartel territory).

During this period, the effects of the war did not occur simultaneously throughout the country. In some places, calm (although initially apparent) remained for several years and then changed. This period of calm during the war was mainly experienced in central areas of the regions occupied by the cartels, where only one group exercised control over the territory.

> In this particular protected area, we only have one criminal group [...] during that time there was no peace, but everything was calm (02-Environmental inspector, Sinaloa Cartel territory).
>
> When I was an inspector in that protected area, things got ugly [dangerous] until 2008, when La Familia clashed with Los Zetas in the town [...] There were decapitated bodies everywhere. It was unbelievable to see the heads of the "malandros" [drug traffickers] in the town square. From that year on, we saw horrible things (14-Environmental inspector, Los Zetas Cartel territory).

As a result of the war, the boundaries between the cartels became blurred as the army and police conducted strategic operations to dismantle the cartels and capture their leaders. In the collective imaginary, the war is thought to have reached its peak between 2008 and 2010, when the country experienced the highest levels of violence in its history (until then).

By 2008, 14,006 people had died as a result of the war, rising to 25 757 deaths in 2010, and ending Felipe Calderón's administration with approximately 70 000 deaths, although the figures have not been fully explained and clarified for the entire Calderón administration (see Bastidas, 2012; Rosen & Zepeda, 2015b).

Environmental conservation activities in protected areas were severely affected during this period. In addition to drug destruction operations, drug laboratories and camps where sicarios and drug crop guards hid posed serious problems for CONANP's environmental conservation activities.

> By 2010, all conservation activities, community projects, surveillance, and biological monitoring had been cancelled. It was not until 2011 that surveillance brigades were reestablished, and in 2012 we restarted activities in the communities. It was not until 2013 that we reestablished all activities (05-Director of Protected Area, Tijuana Cartel Territory).
>
> When there were armed clashes, we couldn't do anything, but when we entered the protected area, it was multipurpose. We tried to do everything in one trip, you know, monitor cameras, meet with communities, etc., and then we would return later (13-Director of protected area, Sinaloa Cartel territory).
>
> The worst we experienced was in 2010, when the government arrested a drug leader. You can imagine how things became. It was so dangerous that once, when we were moni-

> toring birds on some islands where they nested, sicarios arrived in their boats, threatened us, took everything we had, and took our boat [...] leaving us there. We spent about five or six hours in the sun waiting for a fisherman to rescue us and take us back to the shore (06-Director of protected area, Tijuana Cartel territory).

An important aspect of the experiences of inspectors and directors is that many experiences remained only in memory and were not documented, although the directors of the protected areas communicated these experiences to CONANP and decisions were made to reduce risks at each site, such as reducing field activities, prioritising office work, or exercising the highest possible caution. However, the conditions of insecurity outweighed all the strategies that could be implemented in the protected areas.

It should be noted that not all protected areas in Mexico experienced the same level of insecurity during the Mexican War on Drugs; it is possible that some were not affected at all. However, during fieldwork, no protected areas were identified that had not been affected.

Personal Safety Versus Environmental Conservation

During Mexico's War on Drugs, environmental inspectors and directors of protected areas had to learn how to carry out conservation activities in dangerous areas, as their professional or work training did not prepare them to work in violent contexts. Although many experiences were recounted during the interviews, the war had a major impact on conservation efforts in the country's protected areas and posed a constant threat to personal safety.

> We were never taught at university how to act or respond in situations like this. We have learned everything from our own experience and the experiences of other colleagues (04-Director of protected area, territory of the Sinaloa Cartel).
>
> This is the reality of working in protected areas in rural zones: improvising when necessary to ensure our own safety and that of our colleagues, while doing our best to achieve environmental conservation (01-Environmental inspector, Sinaloa Cartel territory).

Although the experiences recounted highlight the dangerous context not only for environmental conservation in protected areas, but also on a personal level for inspectors and directors of protected areas, during this period interactions with drug traffickers did not always involve threats or the loss of work tools. The interviews identified cases in which CONANP personnel suffered physical attacks that violated their personal integrity and rights.

> In 2011, I was on my way to the protected area. That day, I did not want my fellow inspectors to accompany me because I was only going to deliver some bulletins to a community for some activities we had planned to implement. However, before reaching the village, two vans with sicarios blocked my way and forced me out of my van with punches [CONANP vehicles are marked with the institution's logo for identification purposes] and forced me into their truck [...] They beat me and shouted at me to tell them if I had reported them to the Army, but I didn't know what they were talking about. They kept me locked in a room

> for two days with my face covered and handcuffed. They took my phone and called several people to demand a ransom for me. When they realised that no one would pay for my ransom, they beat me and threw me into a wooded area about ten kilometres away. It is a miracle that I am alive, and I am disappointed with everything (12-Environmental inspector, territory of the La Familia Cartel).
>
> In early 2010, two female environmental inspectors were abducted and held captive for a few hours, but we contacted individuals with connections to these groups, people from the community where we work, and they were released. They were left in a remote location, and we went to retrieve them. They were not physically harmed, only frightened. but they never went anywhere alone again (03-Director of protected area, Sinaloa Cartel territory).
>
> The only time I have been close to death was during a shootout. We were arriving at a village in the mountains to work with a community on a reforestation project, but as soon as we entered the village, a fierce shootout broke out between the SEDENA (the army) and the drug traffickers. We got out of the van and took cover as best we could. It was horrible. We had to turn back, terrified. Obviously, the people in the community were not going to participate in the reforestation after the shootout (09-Director of protected area, Gulf Cartel territory).

During this period and even in the post-war period, insecurity in protected areas continued unabated. These experiences, recounted by inspectors and directors of protected areas, seek to highlight a phenomenon that has not been formally documented or considered in academic studies, as was effectively the case in cities, semi-urban towns, and border cities, which received greater media and academic attention.

However, this study shows that although the fundamental principle of the Mexican War on Drugs was to put an end to drug trafficking, in the process, protected legal rights such as property, physical integrity, and the lives of inspectors and directors of protected areas were undermined. This is particularly relevant because these forms of violence in public service can be understood from the perspective of public servants involved in public security issues (police, military, marines, etc.) who are aware of the context of violence due to the nature of their work and know how to respond.

While from the perspective of environmental inspectors and protected area directors these forms of violence have a different connotation, professional profiles such as forest engineers, biologists, environmental engineers, and agroforestry engineers were ideal for environmental conservation, so they were not trained in tactical police or military operations, nor were they armed, and as part of their activities they did not deal with situations of direct violence such as those that occurred during the war period and in the subsequent decades.

In the epistemological corpus of green criminology, crimes can be classified by chromatic areas. This classification was adapted by White (2008) from Crook and Pakulski (1995) and categorises ecological crimes into issues: white, green, and brown; for example, the white issue covers environmental crimes and harms related to technological development, the green issue covers crimes and harms against biodiversity and ecosystems, and the brown issue covers crimes and harms related to urban and industrial life, such as pollution.

However, when analysing the reality of criminal activity in environmental issues in Mexico, in 2023 a new chromatic area was proposed within White's (2008) classification: red issues. This area focuses on analysing crimes against environmental defenders and environmental rights defenders, as well as people dedicated to environmental conservation (see Carpio-Domínguez, 2023; Carpio-Domínguez & Castro-Salazar, 2025).

Attacks on environmental inspectors and directors of protected areas can be considered within this category (red issues), which covers injuries, homicides, and other harm to persons engaged in environmental conservation at all governmental and non-governmental levels (see Wyatt, 2022; Carpio-Domínguez, 2023; Gallegos-Martínez et al., 2023).

The Post-War Period (2012–2024). Lessons Learned and Resilience for Environmental Conservation

With the end of Felipe Calderón's presidential term, the war came to an end, at least in the government's discourse. However, the struggles between cartels for territory, as well as armed clashes with state security forces, continued into the following decade. Discovering that security policies implemented through the capture of drug trafficking leaders was a good political and military strategy, the history of Mexican drug trafficking shows that this strategy does not work (Dickenson, 2014; Rosen & Zepeda, 2015b).

With Enrique Peña Nieto's election as president in December 2012, the Partido Revolucionario Institucional (PRI) returned to the presidency after two terms of the Partido Accion Nacional (PAN), perhaps in an attempt to bring drug trafficking under a single authority, as had been the case for more than 70 years prior to 2000.

However, the public security scenario for 2012 was approximately 70,000 war-related deaths (see Bastidas, 2012; Rosen & Zepeda, 2015b), an increase from seven cartels to 25 criminal groups due to fragmentation caused by the capture of leaders and internal struggles (Fig. 8.3) (see Supplementary Material 8.1), and a sense of widespread social vulnerability at the national level.

This situation of alliances and rivalries between criminal groups contributed to the continuation of armed clashes and conflicts during Peña Nieto's presidential term (2012–2018). Violence caused by fragmentation and clashes between criminal groups continued throughout most of the country.

National policy during this six-year term focused on crime prevention through the "National Programme for the Social Prevention of Violence and Crime", but also on strengthening police and military training, creating the Gendarmerie, and expanding the capacity for mobile deployment through fixed, semi-permanent, and mobile military headquarters to travel to any part of the national territory where it is required and provide citizen security (Kostelnik & Skarbek, 2013; Secretaría de Relaciones Exteriores, 2014; Hernández, 2020).

However, even with the change in the political government and a new security policy, the situation of vulnerability in protected areas did not change for CONANP

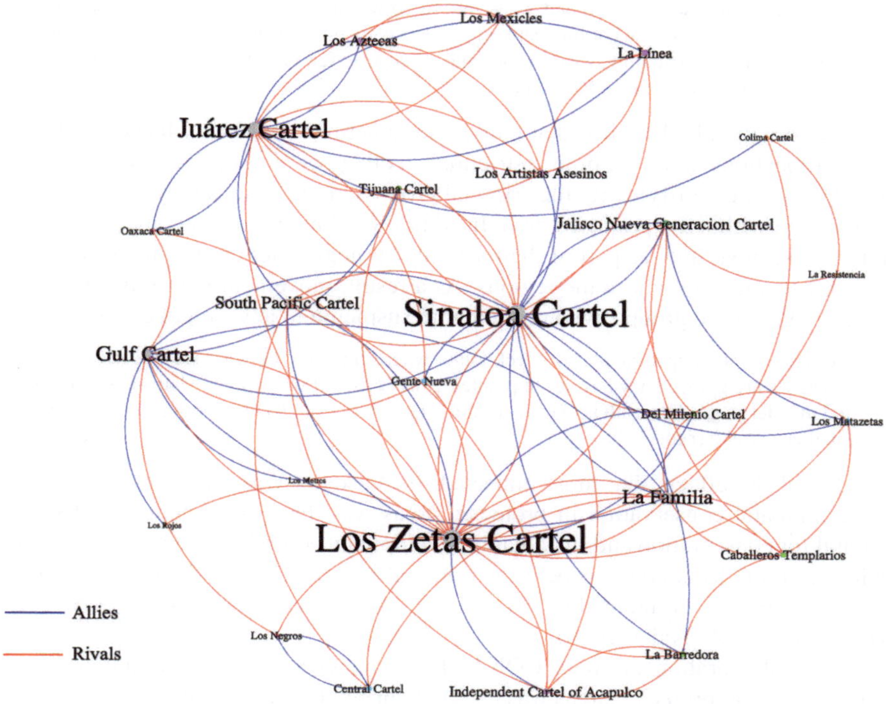

Fig. 8.3 Network of alliances and rivalries between criminal groups in Mexico in early 2012 (post-war period). **Source:** The authors with data from Sin Embargo (2012)

personnel. With a total of 183 protected areas, the presence of criminal groups in protected areas continued to represent the main problem for environmental conservation.

> We cannot ask for support from these institutions [federal police, military or navy] because one day they are here and the next they are gone; they are rotated from place to place, but we are not, so if we did, we would be associated with them and we would not be here talking right now. In addition, we cannot report the presence of these groups in the protected area because there are always people involved with them [other public servants], so we don't take the risk of saying anything. It's better to do what we came here to do as best we can. They do their thing, and we do ours (11-environmental inspector, territory of the Zetas Cartel).
>
> We practically have to ask permission to enter the protected area and do our work, and if anything happens to us, no one will even know (05-Director of protected area, Tijuana Cartel territory).

This vulnerable situation was addressed by protected area personnel using strategies learned during the previous period. In order to implement environmental conservation activities, the communities provided information on security conditions in the protected areas so that personnel could enter and implement environmental activities with certain limitations.

> We already know where to go and where not to go within the protected area, and we also ask the communities how hot [dangerous] the area is to go to work, and if there has been a recent conflict or something, we turn back and ask again a few days later (12-Environmental inspector, territory of the La Familia Cartel).

This situation has implied that conservation actions could only be conducted in certain locations and with time restrictions, as they were not permitted to enter at night or without notifying (to the criminal groups) of the actions they were going to implement within the protected area. In addition, criminal groups constantly monitored these areas and, as part of their security strategies, set up illegal checkpoints, which became common in the regions. As a result, movement within the area was regulated by these groups, who reacted with suspicion to any irregularities.

> We have been stopped at illegal checkpoints where they ask us what we are doing and where we are going. They have even pointed guns at us, but we explain what we are doing and they let us go. The truth is, we are afraid. It is very risky (13-Director of protected area, Sinaloa Cartel territory).

One of the effects that most impacted the Mexican War on Drugs in terms of environmental conservation in protected areas was the loss of human and social capital for conservation and sustainable development activities in these areas, a situation resulting from conflicts between criminal groups and partially documented in research studies (Maldonado, 2011; Ortiz, 2019) and newspaper reports (e.g. Hernández, 2009; Nájar, 2010).

This collaboration between CONANP personnel and the communities living within and near protected areas was affected during the war and post-war period, as many towns and communities were decimated during the armed conflicts, leaving them partially or completely abandoned in various areas of the country.

> The communities participating in the projects we implement within the protected area have been severely affected. There are even ghost communities or villages where all the inhabitants have been displaced or killed. Sometimes we finish projects with a community and when we return months or a year later, there is no one left. This happens because of conflict between criminal groups. If one group was close to a community and then loses its territory to another group, the new group takes revenge on the people. In addition, when one of these groups is near a community, it is common for them to take a cow to eat or, in the worst cases, beat and rape women in the communities, because these people [drug traffickers] come down to the community (01-Environmental inspector, Sinaloa Cartel territory).

Forced migration due to drug trafficking was well documented in several studies during this period (Barrios, 2020; Cruz, 2017; Durin, 2012; Rubio, 2014; Velázquez, 2017). However, forced displacement and attacks against communities within or near protected areas did not receive adequate attention in these studies, possibly due to the inaccessibility of data or because these communities are on the periphery of public interest.

The impact of the fight against drug trafficking and between criminal groups affected not only the social dimension, but also limited the conservation actions of these communities, resulting in considerable losses of social capital and other resources that are essential for environmental conservation and development.

> The saddest thing is that we work for years with a community and suddenly it's gone, they fled, they were killed, or I don't know, but the time, the investment in the projects and everything else is lost, not to mention that it is the communities that support conservation efforts. Everything we do is done with the support of the people (14-Environmental inspector, territory of the Los Zetas cartel).

Even in the post-war context, the relationship between rural communities and narcotics production continued to develop, as these communities have represented the main source of labour for the planting and harvesting of drugs such as marijuana and poppy, and due to the antagonistic relationship with criminal organisations through the emergence of self-defence groups in these communities, as is the case in Michoacán (Rivera, 2014).

> When the people here in the town could no longer tolerate the extortion by the Jalisco New Generation Cartel, several self-defence groups emerged to protect their territories. The people in the community had always worked in drug cultivation, but when they started demanding fees for everything […] for the number of vehicles, for livestock, etc., the people protested, and the only way to protest against these people [drug traffickers] is through firearms (11-environmental inspector, Zetas Cartel territory).

The Emergence of the Environmental Gendarmerie

In this context of vulnerability, the Environmental Gendarmerie was created in 2016 as an initiative that sought to address the conditions of insecurity in protected areas through a collaboration agreement between the Ministry of the Environment and Natural Resources (Secretaria de Medio Ambiente y Recursos Naturales) (SEMARNAT) and the Homeland Security Commission.

Although the implementation of the Environmental Gendarmerie has its origins in the General Law of Ecological Balance (Ley General de Equilibrio Ecológico) (LGEEPA), which establishes that protected areas may request the assistance of security forces to execute inspection activities when other persons obstruct conservation activities in the area (Article 166 of the LGEEPA, 2024), the Environmental Gendarmerie was established as a security force (with police formation) trained in the prevention and response to environmental crimes and administrative infractions. It had specialised equipment (tactical and operational) and the objective of deploying permanently in all protected areas and priority water supply facilities of federal interest to ensure that these are spaces for knowledge generation, recreation, and sustainable development (SEMARNAT, 2016).

This police force was present in 24 protected areas in Mexico and was trained in the protection of ecosystems and endangered species through surveillance, prevention, and response activities against environmental crimes such as illegal logging, deforestation and wildlife trafficking, human invasion of these natural spaces, and administrative infractions in environmental issues (CONANP, 2018; SEMARNAT, 2018).

Although the Environmental Gendarmerie appeared to be an effective public policy for addressing insecurity in protected areas, it left environmental conservation activities feeling more vulnerable. Over the years, CONANP inspectors and directors were able to adapt to these working conditions and forge links with rural and indigenous communities within and near protected areas; the presence of (federal) police figures turned many protected areas into tense spaces for the work of inspectors and directors of protected areas, altering the work dynamics and the trust of rural and indigenous communities in CONANP personnel.

> When the Environmental Gendarmerie arrived, the people [residents and drug traffickers] began to distrust us. They thought we had brought them there [...] Once, a resident of a community asked me: Why did you bring them [the Environmental Gendarmerie] here? You're going to get into trouble with "los malandros" [drug traffickers] (04-Director of protected area, Sinaloa Cartel territory).
>
> "The people" [drug traffickers] sent a person to ask us what the Environmental Gendarmerie were going to do in the area and to tell them how we worked there to avoid problems (06-Director of protected area, Tijuana Cartel territory).

Contrary to intuition, the sense of security that these new police figures might provide never existed for the inspectors and directors of protected areas. However, it should be clarified that the disruption of this "order" of work was not related to any form of participation, silence, or collaboration by CONANP with organised crime, but rather to the power that criminal groups exert over the territories and the people who inhabit them. Furthermore, it should also be clarified that the functions of inspectors and directors of protected areas are environmental conservation, not public security.

> Several communities with whom we worked on sustainable development and environmental conservation activities ceased to receive us with the Environmental Police, and no longer attended the meetings we held with them. How could they attend when they were making money from poppy production? They thought that the Environmental Gendarmerie, with their police uniforms and firearms, were going to arrest them (14-Environmental inspector, territory of the Los Zetas Cartel).
>
> Personally, I think the Environmental Gendarmerie was a good initiative, but poorly implemented. We weren't ready for such a force in protected areas, and I don't think we are ready now [...] at that time, we were safer without them (02-Environmental inspector, Sinaloa Cartel territory).
>
> Many fishermen stopped cooperating with us when the Environmental Gendarmerie were around. During patrols, the fishermen thought we were persecuting them. Personally, I was very hurt by the loss of trust from the communities (05-Director of protected area, Tijuana Cartel territory).

In 2019, with the change of presidential administration, the Environmental Gendarmerie was dissolved, and its police personnel became part of the newly created National Guard and were removed from protected areas. Inspectors and directors of protected areas had to re-establish ties and trust with rural and indigenous communities in order to continue environmental conservation activities.

> When the Environmental Gendarmerie left, we had to regain people's trust and reconnect with them in order to continue our activities. Without the communities, environmental conservation cannot be achieved (11-environmental inspector, territory of the Zetas Cartel).

> The people in the community trusted us, but they did not trust the Environmental Gendarmerie. Perhaps because their only experience with public security comes from shootouts, helicopter chases, and the feeling of vulnerability left behind when they come to deploy operations and then leave (13-Director of protected area, Sinaloa Cartel territory).

Therefore, the Environmental Gendarmerie implemented during Enrique Peña Nieto's six-year presidential term aimed to strengthen security in protected areas. However, it did not take into account 1) the capacity of criminal groups to exercise control over territories, 2) the economic dependence of many rural and indigenous communities on activities related to drug trafficking, stemming from the government's neglect of basic needs and employment opportunities in regions far from urban centres, 3) environmental conservation is conducted with the support of communities, meaning that the presence of the Environmental Gendarmerie affected the trust of communities and criminal groups towards CONANP personnel, and 4) CONANP personnel are not qualified in public security, so their functions are exclusively related to environmental conservation.

Conclusions

The Mexican War on Drugs had a profound impact on social dynamics, perceptions of insecurity, and an escalation of violence at all levels of social life, representing one of the greatest challenges to environmental conservation in Mexico. Although CONANP's conservation actions and the challenges posed by the war were neglected by the public and government and did not receive fair coverage (documented) in the media, it is evidenced that the Mexican War on Drugs had a negative impact on the personal and professional lives of the directors and environmental inspectors of Mexico's protected areas.

From the epistemological corpus of victimology, directors and inspectors of protected areas can be considered direct and indirect victims of the insecurity resulting from the consolidation of drug cartels and the Mexican War on Drugs. They are direct victims due to the attacks and threats that violate their physical integrity and personal safety; in addition, they can be considered indirect victims of public security policies that have not been effective in solving this national problem.

The Mexican War on Drugs exacerbated the phenomenon of the disbanding and consolidation of new criminal groups that not only engaged in drug trafficking but also diversified their activities into other lucrative illegal sectors. In addition, the presence of new criminal groups worsened the working conditions of CONANP personnel as they faced new challenges to environmental conservation, such as the presence of other interests (economic and social) within protected areas and in nearby communities.

The presence of self-defence groups in communities, the monopolisation of the avocado production market in several states in the Pacific Ocean region, corruption among public servants in the security sector, and the diversification and consolidation of activities such as hydrocarbons trafficking, extortion, the disappearance of persons, and incursions into activities such as illegal mining and logging through "legal" sawmills represented only a small part of the challenges to conservation activities in protected areas in Mexico in the following decades.

References

Astorga, L. (2005). *El siglo de las drogas. El narcotráfico, del Porfiriato al nuevo milenio*. Plaza y Janés.

Barrios, M. (2020). Migración por violencia. Dicotomía del desplazamiento interno forzado en Ciudad Juárez. Nexos Observatorio Migrante-Blog de movilidad poblacional. https://migracion.nexos.com.mx/2020/03/migracion-por-violencia-dicotomia-del-desplazamiento-interno-forzado-en-ciudad-juarez/.

Bastidas, S. (2012). México: el sexenio solitario de Felipe Calderón. *Política Exterior, 26*(148), 94–101. http://www.jstor.org/stable/41702701

Capó-Ramírez, G. (2021). La reforma al sistema de justicia penal en México como un instrumento para fortalecer el acceso a la justicia. En Martínez-Solares, V., Gottsbacher, M. and Gutiérrez, M. (Eds.) Miradas locales y problemas regionales de seguridad y justicia en América Latina (pp. 25–52). : Pontificia Universidad Javeriana.

Carpio-Domínguez, J. (2023). *Criminología verde para México: Desarrollo de una perspectiva verde en la herencia criminológica mexicana*. Universidad Autónoma de Tamaulipas-Tirant Lo Blanch. https://doi.org/10.29059/LUAT.317

Carpio-Domínguez, J., & Castro-Salazar, J. (2025). Researching environmental harms in violent environments of the global south. In R. Cavalcanti, D. Fonseca, V. Vegh Weis, K. Carrington, R. Hogg, & J. Scott (Eds.), *The Palgrave handbook of criminology and the global south*. Palgrave Macmillan. https://doi.org/10.1007/978-3-031-74932-2_115-1

Cepeda, F. (2011). *Narcotráfico, financiación política y corrupción*. Ecoe.

Chabat, J. (2014). La política exterior de México: de Calderón a Peña Nieto. En Bonilla, A. y Jaramillo, G. (Eds.) La CELAC en el escenario contemporáneo de América Latina y del Caribe (pp. 27–44). : FLACSO. https://biblio.flacsoandes.edu.ec/libros/digital/53868.pdf#page=26

Chabat, J. (2019). La seguridad en la política exterior de Peña Nieto: el invitado incómodo. *Foro Internacional, 59*(3–4), 989–1020. https://doi.org/10.24201/fi.v59i3-4.2649

Comisión Nacional de Áreas Naturales Protegidas (CONANP). (2018). Grandes logros de la Gendarmería Ambiental en Áreas Naturales Protegidas. SEMARNAT. Gobierno de Mexico. https://www.gob.mx/conanp/prensa/grandes-logros-de-la-gendarmeria-ambiental-en-areas-naturales-protegidas.

Crook, S., & Pakulski, J. (1995). Shades of green: Public opinion on environmental issues in Australia. *Australian Journal of Political Science, 30*(1), 39–55. https://doi.org/10.1080/00323269508402322

Cruz, E. (2017). Movilidad forzada por las dinámicas del narcotráfico: variables para el análisis del proceso de asilo de mexicanos por violencia en EE. UU. *Foro Internacional, 57*(3), 576–606. https://doi.org/10.24201/fi.v52i3.2449

del Olmo, R. (1987). Aerobiology and the war on drugs: A transnational crime. *Crime and Social Justice, 30*, 28–44. https://www.ojp.gov/ncjrs/virtual-library/abstracts/aerobiology-and-war-drugs-transnational-crime

Dickenson, M. (2014). The impact of leadership removal on Mexican drug trafficking organizations. *Journal of Quantitative Criminology, 30*, 651–676. https://doi.org/10.1007/s10940-014-9218-5

Durin, S. (2012). Los que la guerra desplazó: familias del noreste de México en el exilio. *Desacatos, 38*, 29–42. https://www.scielo.org.mx/scielo.php?script=sci_arttext&pid=S1607-050X2012000100003

EFE verde. (2018). El narcotráfico y la tala ilegal aprovechan la protección de áreas de Mesoamérica (22 de noviembre de 2018). EFE verde. https://www.efeverde.com/noticias/narcotrafico-tala-ilegal-areas-protegidas-mesoamerica/.

Estrada, C. (2012). La Iniciativa Mérida y el combate al narcotráfico. Cooperación bajo concepciones inadecuadas. *Revista de El Colegio de San Luis, 2*(3), 266–279. https://doi.org/10.21696/rcsl032012520
Gallegos-Martínez, G., Carpio-Domínguez, J., & Castro-Salazar, I. (2023). Environmental exploitation and violence against indigenous people in Mexico. In D. Goyes (Ed.), *Green crime in the global south. Palgrave studies in green criminology*. Palgrave Macmillan. https://doi.org/10.1007/978-3-031-27754-2_4
Gledhill, J. (2003). Neoliberalismo e ingobernabilidad: caciquis- mo, militarización y movilización popular en el México de Zedillo. Relaciones. Estudios de Historia y Sociedad, 96.
Gobierno de México. (2007). Programa sectorial de seguridad pública (2007–2012). Diario Oficial de la Federación. Gobierno de México. https://www.dof.gob.mx/nota_detalle_popup.php?codigo=5029039.
Goyes, D., & South, N. (2017). Green criminology before "Green Criminology": Amnesia and absences. *Critical Criminology, 25*, 165–181. https://doi.org/10.1007/s10612-017-9357-8
Grillo, I. (2011). *El Narco: inside Mexico's criminal insurgence*. Bloomsbury Press.
Gutiérrez-López, A. (2016). El sistema penal acusatorio en el estado mexicano. *Perfiles De Las Ciencias Sociales, 4*(7) .
Hernández, G. (2020). La política de seguridad en México entre 2012 y 2018. De la confrontación a la prevención del delito. *Espiral, 27*(77), 43–79. https://doi.org/10.32870/eees.v27i77.7082
Hernández, L. (2009). El narcotráfico y la sociedad rural. La Jornada (6 de octubre de 2009). https://www.jornada.com.mx/2009/10/06/opinion/014a1pol.
Hurtado, O., & García-Paz, R. (2013). El narcotráfico en México como problema transnacional. *Revista Mexicana de Política Exterior, 97*, 35–64. https://re.sre.gob.mx/rmpe/index.php/rmpe/article/view/458
Jenner, M. (2011). International drug trafficking: A global problem with a domestic solution. *Indiana Journal of Global Legal Studies, 18*(2), 901–927. https://doi.org/10.2979/indjglolegstu.18.2.901
Kostelnik, J., & Skarbek, D. (2013). Las instituciones de gobernanza de una organización de narcotráfico. *Public Choice, 156*, 95–103. https://doi.org/10.1007/s11127-012-0050-x
La Jornada. (2007). La SEDENA reserva el nombre del químico usado para erradicar plantíos de drogas. 13 de septiembre de 2007. https://www.jornada.com.mx/2007/09/13/index.php?section=politica&article=022n1pol.
Ley Federal contra la Delincuencia Organizada. (2021). Última reforma publicada en el Diario Oficial de la Federación el 25 de mayo de 2021. Cámara de Diputados del H. Congreso de la Unión. Gobierno de México. https://www.diputados.gob.mx/LeyesBiblio/pdf/LFCDO.pdf.
Ley General de Equilibrio Ecológico (LGEEPA). (2024). Última reforma publicada en el Diario Oficial de la Federación el 1 de abril de 2024. Cámara de Diputados del Honorable Congreso de la Unión. Gobierno de México. https://www.diputados.gob.mx/LeyesBiblio/pdf/LGEEPA.pdf.
Malamud, C., & García, C. (2011). El conflicto fronterizo entre Costa Rica y Nicaragua: ¿medio ambiente, soberanía, narcotráfico o mero instrumento electoral? Real Instituto Elcano. http://biblioteca.ribei.org/2045/1/ARI-22-2011.pdf.
Maldonado, S. (2011). Drogas, violencia y militarización en el México rural. El caso de Michoacán. *Revista Mexicana De Sociología, 24*(1), 5–39. http://www.scielo.org.mx/pdf/rms/v74n1/v74n1a1.pdf
Morales, L. (2017). La paz y la protección ambiental en Colombia: propuestas para un desarrollo rural sostenible. http://static.iris.net.co/sostenibilidad/upload/documents/envt-colombia-esp_web-res_final-for-email.pdf.
Moreno-Hernández, M. (2025). La reforma al sistema de justicia penal que México requiere para responder a las exigencias democráticas del Estado de derecho. *Criminalia, 92*(1), 167–199. https://www.criminalia.com.mx/index.php/revista/article/view/253/249
Nájar, A. (2010). *El "apoyo social" del narcotráfico mexicano*. BBC News (14 de diciembre de 2010). https://www.bbc.com/mundo/noticias/2010/12/101213_narcotrafico_mexico_base_social_lav.
Núñez, A. (1999). El narcotráfico en el Perú: estrategias conjuntas de las Fuerzas Armadas y la policía nacional para su erradicación. *Boletín de información, 260*, 79–85. https://dialnet.unirioja.es/servlet/articulo?codigo=4626338

Olinger, M. (2013). La propagación del crimen organizado en Brasil: una mirada a partir de lo ocurrido en la última década. In J. Garzón & E. Olson (Eds.), *La diáspora criminal: la difusión trasnacional del crimen organizado y cómo contener su expansión* (pp. 101–142). Wilson Center. https://www.wilsoncenter.org/sites/default/files/media/documents/publication/LA_DIASPORA_CRIMINAL.pdf

Ortiz, R. (2019). *Cárteles y bandas en la cuenca del Caribe: ¿crimen común o amenaza antiestatal?* IEEE. https://www.ieee.es/Galerias/fichero/docs_investig/2019/DIEEEINV01-2019Carteles.pdf

Rivera, J. (2014). Crimen organizado y autodefensas en México: el caso de Michoacán. : Friedrich Ebert Stiftung. https://library.fes.de/pdf-files/bueros/la-seguridad/10845.pdf.

Rodríguez-Ferreira, O. (2012). *El nuevo sistema de justicia penal acusatorio en México: Análisis descriptivo de la reforma constitucional de* 2008. Trans-Border Institute, Joan B. Kroc School of Peace Studies. University of San Diego. https://catcher.sandiego.edu/items/peacestudies/Cap2_An%C3%A1lisis_Rodr%C3%ADguez_120403(dist).pdf.

Rosen, J., & Zepeda, R. (2015a). La guerra contra el narcotráfico en México: una guerra perdida. *Revista Reflexiones, 94*(1), 153–168. https://www.scielo.sa.cr/pdf/reflexiones/v94n1/1659-2859-reflexiones-94-01-00153.pdf

Rosen, J., & Zepeda, R. (2015b). Una década de narcoviolencia en México: 2006-2016. In R. Benítez-Manaut & S. Quezada (Eds.), *Atlas de la Seguridad y la Defensa de México 2016.* Casede. https://casede.org/PublicacionesCasede/Atlas2016/JonathanD_Rozen_Roberto_Zepeda.pdf

Rubio, L. (2014). *Desplazamiento interno inducido por la violencia: una experiencia global, una realidad mexicana.* ITAM. https://www.cmdpdh.org/publicaciones-pdf/libro_desplazamiento_una_realidad_mexicana.pdf

Secretaria de Medio Ambiente y Recursos Naturales (SEMARNAT). (2016). *Gendarmería Ambiental: más seguridad en las ANP.* SEMARNAT. Gobierno de México. https://www.gob.mx/semarnat/articulos/gendarmeria-ambiental-mas-seguridad-en-las-anp#:~:text=La%20gendarmería%20se%20constituirá%20como,sustentable%20para%20todos%20los%20mexicanos.

Secretaria de Medio Ambiente y Recursos Naturales (SEMARNAT). (2018). Acciones ambientales la creación de la Gendarmería Ambiental. SEMARNAT. Gobierno de México. https://www.gob.mx/semarnat/articulos/accionesambientales-creacion-de-la-gendarmeria-ambiental-172050.

Secretaría de Relaciones Exteriores. (2014). Enrique Peña Nieto presenta la nueva división de Gendarmería de la Policía Federal. *Boletín Informativo "La Política de Derechos Humanos de México", 70*, 1–2. https://embamex.sre.gob.mx/eua/images/pdf/boletines/023_Boletn70.pdf

Sin Embargo. (2012). *De siete al inicio del sexenio, los cárteles mexicanos se multiplicaron: Calderón dejará por lo menos 25.* Sin Embargo (August 20, 2012). México. https://www.sinembargo.mx/335244/de-siete-al-inicio-del-sexenio-los-carteles-mexicanos-se-multiplicaron-calderon-dejara-por-lo-menos-25/.

United Nations. (2000a). United Nations Convention against Transnational Organized Crime. General Assembly resolution 55/25 of 15 November 2000. Vienna: UN. https://www.unodc.org/unodc/en/organized-crime/intro/UNTOC.html.

United Nations. (2000b). *United nations convention against transnational organized crime.* General Assembly resolution 55/25, adopted on 15 November 2000. Mexico: signed on 13 December 2000; ratified on 4 March 2003. https://treaties.un.org.

United States Environmental ProtectionAgency (EPA). (2024). Paraquat Dichloride. US Government. https://www.epa.gov/ingredients-used-pesticide-products/paraquat-dichloride.

Valdés, G. (2013). *Historia del narcotráfico en México.* Editorial Aguilar.

Velázquez, A. (2017). *Desplazamiento interno por violencia en México. Causas, consecuencias y responsabilidades.* CNDH. http://www.corteidh.or.cr/tablas/r37820.pdf

Wyatt, T. (2022). Wildlife trafficking. A deconstruction of the crime, victims and offenders. : Palgrave Macmillan. https://doi.org/10.1007/978-3-030-83753-2.

White, R. (2008). *Crimes against nature: Environmental criminology and ecological justice.* Willan Publishing. https://doi.org/10.4324/9781315880723

Part III
Convergence of Drug Trafficking in Natural Protected Areas in 2024–2025

Overview

This section analyses the current challenges for environmental conservation in protected areas considering their convergence with drug trafficking territories and the environmental, social, and public security impacts of this convergence. The analysis in this section considers the changes in the public agendas of recent presidential administrations.

These chapters analyse the geographical distribution of drug trafficking groups in Protected Natural Areas and how inspectors and public servants of the Comisión Nacional de Áreas Naturales Protegidas (National Commission of Protected Natural Areas) and the Procuraduría Federal de Protección al Ambiente (Federal Environmental Protection Agency) face the challenges for environmental protection and conservation. It also explores the new (i)legal economic markets and criminal activities that criminal groups have implemented within and near protected areas as well as their social and environmental impacts in various regions of the country.

Chapter 9 provides a detailed analysis of how drug trafficking organisations utilise protected areas for production, processing, and trafficking, and the patterns of their geographical distribution across the country. It also provides an overview of the current public security conditions in the protected areas and offers important information to guide public policy decision-making through scientific information.

Chapter 10 analyses the "new" economic and illegal activities that criminal groups operate within and near protected areas. The study of these activities is important to understand the adaptation of criminal groups to "new" economic markets as a strategy for diversification of activities, survival, and permanence even against government efforts to combat these criminal groups in the country.

Chapter 9
Geographical Distribution of Drug Trafficking in Protected Areas

Introduction

Enrique Peña Nieto's presidential term in 2018 marked the end of direct action against drug trafficking in Mexico. Although not officially recognised in this statement, the set of public policies on security during the last two presidential terms in Mexico (2006–2012 and 2012–2018) focused on addressing the issue through direct confrontation with drug trafficking.

During Peña Nieto's presidential term, a list of 122 criminals linked to drug trafficking in Mexico was compiled. This policy led to the neutralisation of 110 of them, of whom 96 were arrested and 14 were killed or murdered in armed confrontations with the security forces (see Supplementary Material 9.1). This list represented one of the priority public security issues within the Peña Nieto administration and determined the public security scenario for the subsequent 6-year terms.

According to estimates, the drug trafficking conflict between 2006 and 2018 left a total of 148 criminal groups nationwide. However, although the total number has not been accurately documented, sources report that by the end of 2024 there were 175 criminal groups in Mexico (Roa, 2020). Furthermore, this variability in criminal groups has led to the diversification of legal and illegal economic activities that have allowed them to remain in business over time and in the competition amongst illegal markets. However, drug trafficking has continued to be the most profitable illegal activity for criminal groups in Mexico.

The end of this 6-year term and the social impact on public security led to widespread dissatisfaction with the effectiveness of government institutions in tackling crime, resulting in a change of political direction for the following two six-year presidential terms. With the entry into the presidency of the Movimiento Regeneración Nacional (MORENA) (National Regeneration Movement) political

---

Supplementary Information The online version contains supplementary material available at https://doi.org/10.1007/978-3-032-18218-0_9.

J. L. Carpio-Domínguez et al., *Drug Trafficking and Protected Areas in Mexico*, https://doi.org/10.1007/978-3-032-18218-0_9

party in 2018, with a left-wing political orientation, the alternation between the neoliberal and centre-left Partido Revolucionario Institucional (PRI) (Institutional Revolutionary Party) (in the last six-year term) and the centre-right and right-wing Partido Acción Nacional (PAN) (National Action Party) came to an end.

Meanwhile, in the United States, Donald Trump won the presidential election for the 2017–2021 term with a critical stance and a heavy-handed approach to Mexican cartels. With a nationalist focus, he considered Mexican cartels a threat to homeland security and promoted border closure through the construction of a wall to prevent the entry of migrants and drugs. He also accused the Mexican government of failing to control drug trafficking and claimed that, as a result, the United States was facing a national emergency due to the opioid crisis. He declared Mexican cartels to be Foreign Terrorist Organisations (FTOs), primarily the Jalisco Nueva Generación Cartel and the Sinaloa Cartel (Morales-Gámez and Ortega-Ramírez, 2024).

In Mexico, after 18 years of environmental conservation in violent and vulnerable conditions, the personnel of the National Commission of Natural Protected Areas (CONANP) have developed the experience and resilience to implement environmental conservation activities in collaboration with rural and indigenous communities within and near protected areas, despite the strengthening of criminal activities by at least 175 criminal groups nationwide (Cano, 2024; Roa, 2020) that not only engage in drug trafficking but have also diversified their illegal activities into other sources of economic income alternative to drug trafficking.

This chapter revisits the experiences of CONANP directors and inspectors to highlight the working conditions for environmental conservation in Mexico's protected areas. It also includes a documentary analysis that provides a historical overview of the main events related to public security and environmental conservation in Mexico.

This chapter reviews the main public security and environmental conservation policies during the presidential term of Andrés Manuel López Obrador (2018–2024), with policies of non-confrontation with drug trafficking under the slogan "hugs, not bullets", as well as the first year of Claudia Sheinbaum Pardo's presidential term (2024–2025). It should be clarified that this chapter is not intended to be a historiographical analysis of drug trafficking in Mexico during this period, but rather to highlight historical events in order to document the environmental conservation experiences of CONANP personnel in the current context of Mexico, focusing on the presence of organised crime within protected areas.

The Policy of "Hugs, Not Bullets" (2018–2024)

With Andrés Manuel López Obrador's presidential election in Mexico, public security policy changed at the national level. With a discourse of non-confrontation with criminal groups, the 2018–2024 National Peace and Security Plan aimed to rethink national security and reorient the Armed Forces (army and navy) and create the National Guard, eliminating the federal police and removing the military from the streets in order to address the situation of public insecurity in Mexico (Serrano-Carreto, 2019).

Using political rhetoric to denounce the insecurity caused by the wars waged by the two previous presidents (mainly Felipe Calderón Hinojosa), Andrés Manuel López Obrador's policy focused on consolidating social support programmes as a (primary) crime prevention strategy under the slogan "hugs, not bullets". Although the slogan is not official, it is attributed to the characteristics of non-direct confrontation with criminal groups during his presidential term.

To address all scenarios of insecurity and crime in Mexico, the 2018–2024 National Peace and Security Strategy focused on addressing the country's structural problems, mainly by: a) eradicating corruption and reactivating the administration of justice, b) guaranteeing employment, education, and health, c) guaranteeing respect for and promotion of human rights, d) regenerating society's ethics, e) reformulating the fight against drugs, f) undertaking peacebuilding, and g) regaining control of prisons and restoring their dignity (Gobierno de México, 2018; Secretaria de Seguridad y Protección Ciudadana, 2018).

Although the actual figures for the total number of criminal groups have not been officially confirmed, in 2018, there were an estimated 148 criminal groups distributed nationwide, with a notable territorial expansion of the Jalisco Nueva Generacion Cartel (CJNG), which had a presence in 31 of Mexico's 32 states, followed by the Sinaloa (Pacific) Cartel, which had a presence in 16 states (Roa, 2020). After Andrés Manuel López Obrador's presidential term, there were an estimated 175 criminal groups (Cano, 2024).

The insecurity during Andrés Manuel López Obrador's six-year term was characterised by the consolidation of illegal activities such as hydrocarbon trafficking (huachicol) (involving the military, navy, politicians, businesspeople, and criminal groups), alliances between criminal groups and politicians to administer extortion ("cobro de piso" or protection fee) for public transport, extortion for the sale of primary products (mainly agricultural and livestock), electoral fraud, lack of accountability for public spending, budget cuts to government institutions, the disappearance of autonomous institutions, the intention to exercise control over the three Powers of the Union (Executive [Presidency], Legislative [Deputies and Senators], and Judicial [Supreme Court of Justice of the Nation and Prosecutors' Offices]), as well as the implementation of public policies and government programmes whose actual rate of success or failure is unknown (see Alvarado-Mendoza, 2024).

In this scenario, one of the main public policies in public security was the creation of the National Guard in 2019 as part of the strategies to "remove" the military and marines from the streets. The creation of the National Guard was initially an attempt to create a non-militarised security force that would not only meet public security needs but also permeate the country's weakened federal police institutions. However, in practice, it ended up being made up of military personnel, air force personnel, and trusted civilian personnel with police training (Article 4 of the Ley de la Guardia Nacional, 2025).

The objective of the National Guard is to perform the public security function under the responsibility of the Federation and, where appropriate, to temporarily collaborate in public security tasks that correspond to the federal entities or municipalities, with the aim of preserving social peace and protecting the citizens' rights (Article 5 of the Ley de la Guardia Nacional, 2025).

Among the main criticisms of this police force is that more than half of its officers are from the military and navy, which is a departure from the original idea of a civilian public security forcé (see Instituto Nacional de Estadística y Geografía [INEGI], 2024). However, among the main phenomena addressed by the National Guard are those related to drug trafficking and organised crime through arrests and seizures of firearms and drugs such as marijuana, cocaine, fentanyl, lysergic acid, and other derivatives of D-lysergic acid.

It should be noted that among the activities conducted by the National Guard are seizures of clandestine hydrocarbon taps (fuel theft or huachicol), with up to 2 745 clandestine taps seized in 2023 (INEGI, 2024). Although this phenomenon initially appeared to be random among the criminal activities in the country, during this six-year term it became one of the main sources of financial income for criminal groups through the formation of international criminal networks with the participation of federal government personnel, the Navy, and organised crime.

Furthermore, in the absence of the "Environmental Gendarmerie" proposed during the previous six-year term, the National Guard is authorised to participate in coordination with environmental authorities in the protection of natural resources, as well as in the prevention, investigation, and combating of environmental crimes that pose a risk to public safety, including illegal logging, wildlife trafficking, and the degradation or contamination of ecosystems. Its actions must be guided by the principles of respect for human rights, environmental protection, and inter-institutional collaboration (Article 9, Section 25 of the Ley de la Guardia Nacional, 2025).

In summary, this period started with challenges on several fronts in public service, for example, an increase in criminal groups, increased violence, diversification of illegal activities by organised crime, social expectations following the alternation of two political parties, the main focus in the discourse on strategies to combat poverty and defend the nation from foreign influences, to mention only some of them (see Gobierno de México, 2018).

Geographical Sectorisation of Public Security in Mexico

One of the strategies implemented in public security at the national level has been the sectorisation of the national territory. This strategy seeks to identify common social problems, optimise the allocation of police resources, and optimise the government's response. This strategy was formally implemented in previous six-year terms (see Secretaría de Gobernación, 2013; Montalvo, 2013) and continued in subsequent six-year terms (Tovar, 2021).

Fig. 9.1 Public security sectorisation in Mexico by region. **Source:** the authors with information from Secretaría de Gobernación (2013)

The country is divided into five regions: Northwest, Northeast, West, Central, and Southeast. The western region has the largest number of states and is interconnected with the main regions where criminal groups and drug cartels are concentrated, as well as with emerging and established organised crime groups until this period of time (Fig. 9.1).

In addition to being a strategy for implementing public policies, this sectorisation will serve in this study to highlight common problems in the states of each region, as well as to effectively highlight the presence of criminal groups, protected areas, and the various phenomena of interest arising from the presence of criminal groups in Mexico's regions and protected areas.

Reduction in Federal Budget for Protected Areas

During the first years of Andrés Manuel López Obrador's six-year term, there was a centralisation of power in the country. Presidential decrees abolished autonomous institutions such as the Federal Economic Competition Commission (COFECE), the Energy Regulatory Commission (CRE), the National Hydrocarbons Commission (CNH), the National Council for the Evaluation of Social Development Policy (CONEVAL), the National Commission for the Continuous Improvement of

Education (MEJOREDU), and even the National Institute for Transparency, Access to Information and Protection of Personal Data (INAI) (Confederación Patronal de la República Mexicana, 2024), among others, as well as the mass dismissal of workers and threats to abolish environmental institutions such as the Commission for the Use and Knowledge of Biodiversity (CONABIO) (Hernández, 2023), and put institutions at a disadvantage in addressing current challenges that require measurements and evaluations of the effectiveness of public policies planned and implemented.

In addition to widespread public insecurity, protected areas faced both old and new challenges during this government term. Among the new challenges are federal budget cuts, with only 10 Mexican pesos per hectare of protected area (0.54 US dollars) being allocated (Centro Mexicano de Derecho Ambiental [CEMDA], 2023a), which has exacerbated previous challenges such as the lack of financial resources for equipment.

Between 2008 and 2025, the federal budget allocated to CONANP (2025) has varied with a downward trend, falling from 771 million Mexican pesos (approximately 41.7 million US dollars) in 2008 to 1 001 million Mexican pesos in 2025 (54.2 million US dollars). From 2008 to 2016, there were increases in the budget allocated to environmental conservation activities in protected areas, with 2016 being the year in which the most money was allocated to CONANP, with 1 359 million Mexican pesos (73.5 million US dollars). However, starting in 2017, there was an almost constant downward trend in revenue, with 1 098 million Mexican pesos (59.4 million US dollars), and a slight increase in 2018 with 1 132 million Mexican pesos (61.2 million US dollars) (Fig. 9.2).

Despite the budget reduction in 2017, it was in that year that the highest percentage of the total budget allocated to the environment and natural resources was assigned to CONANP, 3.05% of the total allocated to the Secretaria de Medio

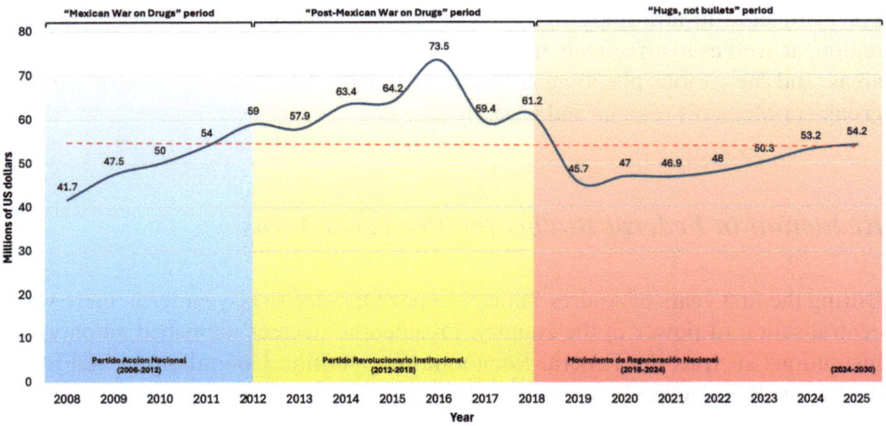

Fig. 9.2 Federal budget allocated to protected areas by presidential term-political party. **Source:** the authors with data from Secretaría de Hacienda y Crédito Público (SHCP) (2025)

Ambiente y Recursos Naturales (SEMARNAT) (Secretariat of the Environment and Natural Resources). On average, between 2008 and 2025, 2.2% of the total government budget allocated to issues related to the environment and natural resources is allocated to Comisión Nacional de Areas Naturales Protegidas (CONANP) (National Commission for Protected Areas); even so, it is one of the environmental institutions that receives the least financial resources.

The lack of federal resources allocated to environmental conservation through CONANP not only limits the institution's personnel's ability to act, but also permeates other aspects such as job security, staff adequacy, and the government's lack of interest in environmental conservation and the country's natural capital, as well as the institution's personnel and the work they do with indigenous and rural communities near and within protected areas.

The criteria of "republican austerity" implemented in many government institutions during the six-year term do not justify the reduction in the budget for environmental conservation activities, since among the main effects of these cuts are the inability to purchase forest firefighting equipment, transport vehicles within protected areas, uniforms, and other basic work equipment that is fundamental and essential for environmental conservation work in Mexico's protected areas.

According to figures from México Evalúa (2025), public spending during 2018–2024 was concentrated on social protection assistance programmes, increasing by 68% compared to the previous six-year period, while areas such as science and technology (−10%), agriculture (−21%), and environmental conservation (−21%) saw decreases in resource allocation despite the importance of these sectors for the country's development.

Furthermore, areas such as homeland security received only 0.8% of gross domestic product (GDP), representing stagnation in federal budget allocation since 2011, despite being one of the most important issues for the Mexican population due to social vulnerability, violence, and social neglect by government institutions (México Evalúa, 2025).

In summary, the distribution of federal economic resources at the national level during this period limited the efforts of government institutions in environmental issues to respond to current challenges not only in terms of public safety within protected areas, but also in the administration of economic resources by Mexico's environmental institutions.

Organised Crime Scenario in Mexico Until 2024

Although drug trafficking dominates the organised crime scene in Mexico, criminal groups are no longer solely involved in drug trafficking. The diversification of illegal and legal activities (often for money laundering purposes) has shaped the drug trafficking scenario. The virus of "generating money wherever possible" implemented by the Los Zetas Cartel influenced the variability of criminal activities within cartels and criminal groups in the following decades until the present day.

Several studies in Mexico have documented the diversification of organised crime activities in recent decades (see Buscaglia, 2015; Cumplido-Tercero, 2015; Fuentes-Díaz, 2015; Vázquez Vadéz, 2018). This diversification included not only drug trafficking, but also kidnapping, extortion, protection rackets, and various forms of illegal trafficking, such as firearms, people, minerals, hydrocarbons, and wildlife (Carpio-Domínguez et al., 2022). In addition to legal activities related to or financed by illegal money, such as currency exchange offices, sawmills, petrol stations (for money laundering and hydrocarbon laundering), alcoholic beverage outlets, and the incursion of members of criminal groups (or those linked to criminal groups) into the political arena as deputies, senators, municipal presidents, etc.

Furthermore, at the end of Peña Nieto's six-year presidential term (2012–2018) and at the beginning of Andrés Manuel López Obrador's six-year administration (2018–2024), fentanyl emerged as a new drug on the Mexican drug trafficking scene, with negative socio-political and security impacts in Mexico and the United States, adding to marijuana, cocaine, heroin, and methamphetamines (Jones & Alonso-Ramos, 2025). The fentanyl crisis led to more than 70,000 deaths of American citizens in 2019, and since 2020, more than 100,000 deaths per year (National Vital Statistics System, 2025).

President Donald Trump's (re)election to the presidency of the United States in 2025 also marked a new phase in the fight against drug trafficking in North America. The United States continued to refer to Mexican cartels as Foreign Terrorist Organisations (FTOs) (Morales-Gámez & Ortega-Ramírez, 2024), and United States authorities even claimed that Mexican criminal organisations had begun to produce fentanyl by importing the precursors for its preparation from China and India (Barr, 2022).

However, in Mexico, organised crime intensified its violent activities in the territories it controls during this period. In addition, other criminal groups dedicated to extortion ("cobro de piso" protection fees) and hydrocarbon theft ("huachicol") consolidated their position as leaders of these criminal activities through violence and government corruption. Among the criminal groups that managed to strengthen their position and become the most violent cartel with the largest territory under their control at the national level was the Jalisco Nueva Generación Cartel (CJNG), which even surpassed the Sinaloa (Pacific) Cartel.

To illustrate the organised crime scenario in Mexico during this period, data on the distribution of criminal groups in municipalities with nationally protected areas was analysed. Until 2024, 82 criminal groups had been identified in 376 municipalities that wholly or partially cover a protected area in Mexico (see Supplementary Material 9.2). These 82 criminal groups represent 48.5% of the 175 criminal groups identified at the national level. The fragmentation of cartels into smaller criminal groups also influenced the legal and illegal activities they perform to obtain economic gains and finance each criminal group.

The Jalisco Nueva Generación Cartel (CJNG) was the criminal group with the largest presence in municipalities with protected areas, at 25.8% (n = 162 municipalities), followed by the Sinaloa Cartel (11.0%, n = 69) and the Gulf Cartel (7.0%, n = 44) (Fig. 9.3). Furthermore, these three criminal groups are characterised by

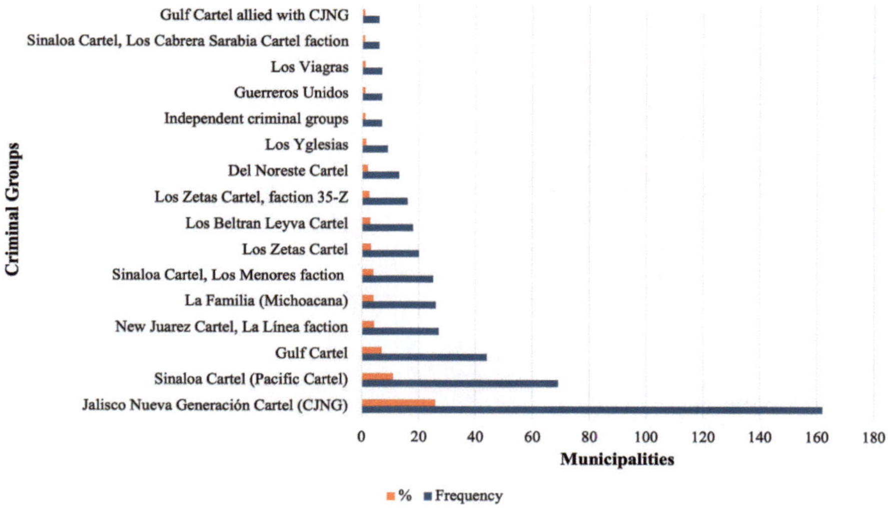

Fig. 9.3 Main criminal groups in municipalities with protected areas in Mexico until 2024. Source: The authors with data from El Universal (2025). Note: to view the total number of groups, please refer to the Supplementary Material 9.2

having the largest territory of influence, not only in protected areas, but throughout the country.

Public insecurity during the 2018–2024 six-year term has been exacerbated, and insecurity caused by criminal groups has not only been represented by drug trafficking, but also by extortion, kidnapping, deprivation of liberty, disappeared persons, protection fees, and various forms of illegal trafficking, such as firearms, persons, minerals, hydrocarbons, and wildlife, as well as an increase in homicides ($n = 244{,}599$) and disappeared persons ($n = 134{,}458$), the highest levels in the country's history (see Comisión Nacional de Búsqueda, 2025; Instituto Nacional de Estadística y Geografía [INEGI], 2025).

Until 2024, of the total number of municipalities that compose the Mexican territory ($n = 2478$), 26.0% ($n = 645$) have at least one protected area, and in 58.2% ($n = 376$) of these municipalities with protected areas, the presence of at least one organised crime group was identified (Fig. 9.4). The Northeast and Northwest regions of the country are noteworthy because most municipalities with protected areas have organised crime groups; in the Northwest, 92.0% of municipalities with protected areas have criminal groups, and in the Northeast, approximately 86.8% have organised crime groups.

In 2024, 82 different organised crime groups were documented within the country's municipalities with protected areas. When considering the distribution of all criminal groups, a total of 627 alliances between criminal groups distributed within municipalities with protected areas were counted; this is due to the fact that some municipalities have more than one criminal group.

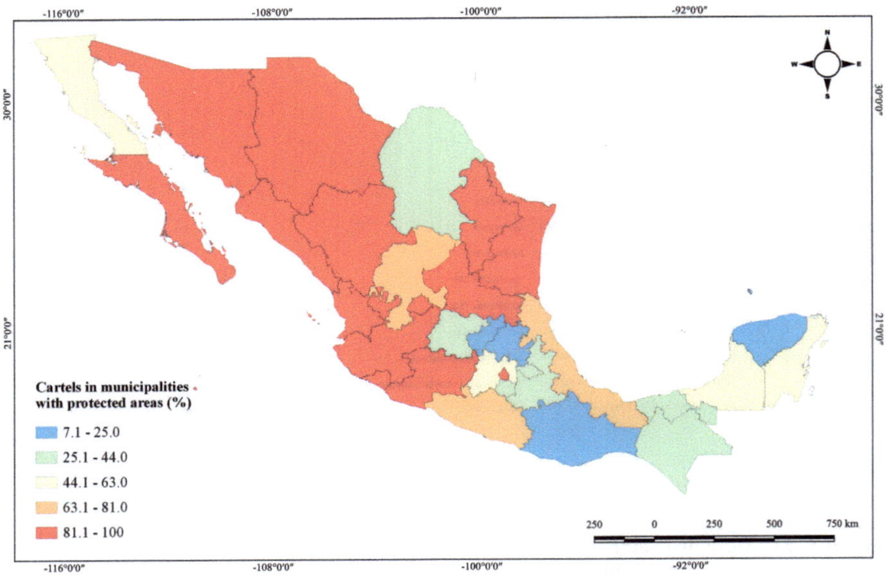

Fig. 9.4 Percentage of criminal groups in municipalities with protected areas in Mexico until 2024. **Source:** The authors with data from El Universal (2025) and Commission for the Use and Knowledge of Biodiversity (CONABIO) (2023)

Organised Crime in Mexico's Protected Areas in 2025

Until 2025, Mexico has 232 protected areas and 175 criminal groups nationwide. It was identified that criminal groups are strategically distributed throughout the country's protected areas (Fig. 9.5), considering the orographic and natural characteristics of each protected area to conduct various activities such as drug cultivation, training camps for sicarios, refuges to evade security forces, the installation of narco-laboratories, and more recently, as ideal locations for clandestine graves.

The presence of criminal groups throughout the country led to an adaptation to new opportunities for legal and illegal financial gain. Large sums of money allowed drug traffickers to establish shops, businesses, and mainly agricultural and livestock activities, which, on the one hand, serve to launder drug money, but also allow them to monopolise these markets at the local level, leading to a reduction in competition among agricultural and livestock producers.

In addition, several areas in the western region have established themselves as leaders in poppy production, even surpassing the Golden Triangle in the northeast of the country. In states such as Guerrero and Michoacán, poppy cultivation in mountainous areas is the main source of income for indigenous and rural communities in the region. Even the dynamics of using residents as labour for drug production remain historically the same as in the Golden Triangle region: a) the criminal

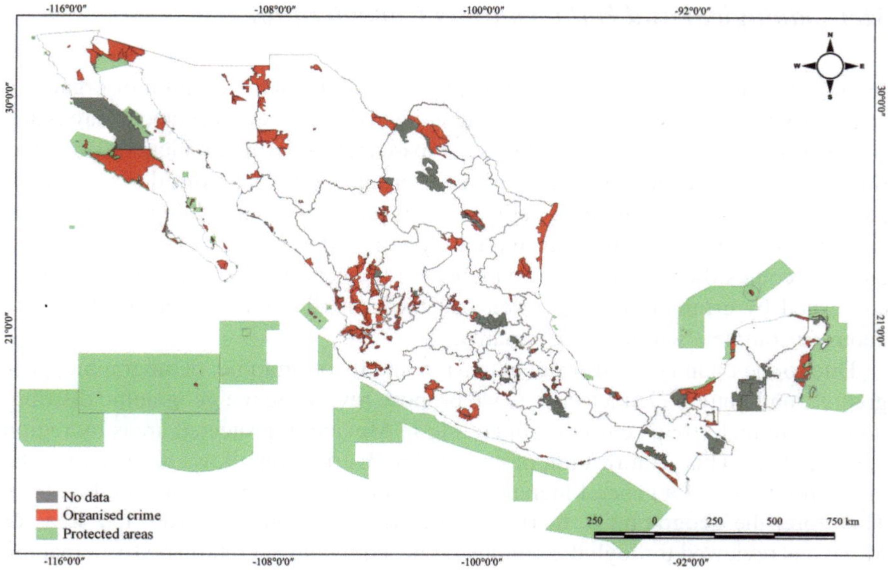

Fig. 9.5 Geographical distribution of drug cartels in protected areas of Mexico until 2024. **Source:** The authors with data from El Universal (2025) and CONABIO (2023). Note: The map shows only the municipalities where criminal groups converge in protected areas. The dominance of criminal groups certainly extends to other municipalities and states

group arrives in the community, b) they ask them to plant marijuana or poppies, c) they offer to buy the entire production and, as a guarantee, give them an advance payment at that time. In addition, they provide courses for the communities on how to cultivate and harvest.

For almost everyone in the communities in Mexico where drugs are cultivated, this activity represents an opportunity to earn enough money to survive, pay debts, acquire goods, etc., which they would not be able to do with traditional agricultural work (corn, beans, oats, etc.). However, drug cultivation is not the only illegal activity by organised crime in protected areas. The diversification and incursion of drug trafficking into many legal and illegal markets has been one of the main limitations not only for its study but also for the development of public policies to tackle organised crime in Mexico.

Criminal Groups in Protected Areas by Region in Mexico

In general, the geographical distribution of natural areas is concentrated as follows: the Southeast region has the highest number of protected areas ($n = 91$), followed by the Central region ($n = 51$), West ($n = 51$), Northwest ($n = 51$), and Northeast ($n = 28$).

Methodological and Data Analysis Considerations

In order to understand, at least partially, the geographical distribution of organised crime groups in protected areas, the geographical distribution of protected areas and current criminal groups in Mexico was compared, considering municipalities as territories where protected areas and criminal groups coincide. Considering municipalities as units of measurement in this study is purely for methodological purposes, since the use of geographic information systems allows for practical comparison and overcomes the limitations of data access to accurately identify the territories controlled by criminal groups, given that their distribution is often not homogeneous or complete within a municipality.

Due to the country's territorial extension and its orographic, demographic, geographic, and natural characteristics, we propose reviewing the geographic distribution of criminal groups currently operating in Mexico in protected areas by region (see Fig. 9.1). This sectorisation of geographic distribution allows to overcome the visual limitation of a general map that considers all this information (see Fig. 9.5). Therefore, the geographical distribution of criminal groups in protected areas of Mexico is reviewed through the regions proposed by the Secretaría de Gobernación (2013): West, Centre, Northwest, Southeast, and Northeast.

Another limitation that must be considered is the lack of currently updated information regarding criminal groups operating within the national territory. Although this chapter covers most of them, it is possible that new criminal groups may have emerged by the time this book is published, or that some may have ceased to exist due to internal conflicts, having been absorbed into another larger criminal group, or having disappeared entirely as a result of territorial disputes.

Some names of criminal groups were abbreviated because some are too long to fit on the maps, mainly when there are associations and alliances between cartels. Table 9.1 shows the names of criminal groups that were abbreviated on the maps for better understanding.

The mapping prioritised the use of the names of criminal groups in Spanish for two reasons: a) to maintain the original name by which they are known and identified in Mexico and Latin America, and b) to facilitate their identification in academic literature, the media, and official government reports.

Table 9.1 Abbreviations of criminal group names in the mapping by region

Abbreviation	Spanish name	Literal translation (English)
CJNG	Cartel de Jalisco Nueva Generación	The Jalisco New Generation Cartel
ZVE	Los Zetas Vieja Escuela	The Old School Zetas Cartel ("originals")
CDG	Cartel del Golfo	The Gulf Cartel
CDN	Cartel del Noreste	The Northeast Cartel
CSRL	Cartel de Santa Rosa de Lima	The Santa Rosa de Lima Cartel

Note: The names of criminal groups are retained in Spanish on the maps, except for those already identified in English in the literature

Drug Trafficking in Protected Areas of the West Region

The western region is located on the Pacific Ocean coast and comprises the states of Aguascalientes, Colima, Guanajuato, Jalisco, Michoacán, Nayarit, Querétaro, and Zacatecas. This region has the highest concentration of cartels and criminal groups distributed across municipalities with protected areas (n = 167 municipalities with at least one criminal group).

In the western region, Jalisco (n = 12), Michoacán (n = 12), and Nayarit (n = 7) are the states with the most protected areas, which are mainly Biosphere Reserves (n = 13), National Parks (n = 13), and Sanctuaries (n = 9).

Although organisations involved in narcotics production had existed in this region since the 1940s, this geographical area has played one of the most significant roles in the history of drug trafficking in Mexico.

In municipalities with protected areas in the western region of the country, 17 different criminal groups with varying levels of association were identified, distributed across 167 municipalities. The three cartels with the strongest presence in that region were: Jalisco Nueva Generacion Cartel (CJNG) (n = 104 municipalities), Sinaloa (Pacific) Cartel (n = 13), and Gulf Cartel (n = 9) (Fig. 9.6).

The western region has historically been characterised by its location in one of the most conflictive areas in the country's recent history, known as "Tierra Caliente" (Hot Land). This area, located in the south of the state of Michoacán, the south of the state of Guerrero and the south of the state of Mexico, is characterised by its high temperatures, which favour agricultural production.

This area is notable for being the geographical, historical, and social space from which the diversification of illegal economic activities by drug cartels such as the Gulf Cartel, through its armed faction Los Zetas, was traced during the 2000s (Grillo, 2011; Valdés, 2013). Los Zetas' operations in the region included extortion, kidnapping, control of mines, crop marketing and agricultural monopoly, illegal logging, and drug production and trafficking (Fuentes-Díaz, 2015).

From poppy and marijuana production to the trafficking of cocaine from Colombia, which is landed on the coasts of Michoacán, Guerrero, and Jalisco, and the establishment of Los Zetas and their fight against and extermination of the Milenio Cartel, this region has become strategically important in the Mexican drug trade. In addition, the first self-defence groups, such as La Familia Michoacana, emerged in this region to confront the extortion practices of Los Zetas and as a response to the government neglect that had characterised this rural area for many years (Fuentes-Díaz, 2015; Maldonado-Aranda, 2012, 2014).

Although describing and analysing the history of drug trafficking and self-defence groups in the region would exceed the length of this study, various studies in Mexico have documented not only the historical emergence of these groups but also the governmental, political, economic, and security changes in the region (e.g. Hernández-Soc, 2019; Hernández-Morales, 2023; López-León, 2017; Maldonado-Aranda, 2012; Mendoza, 2024).

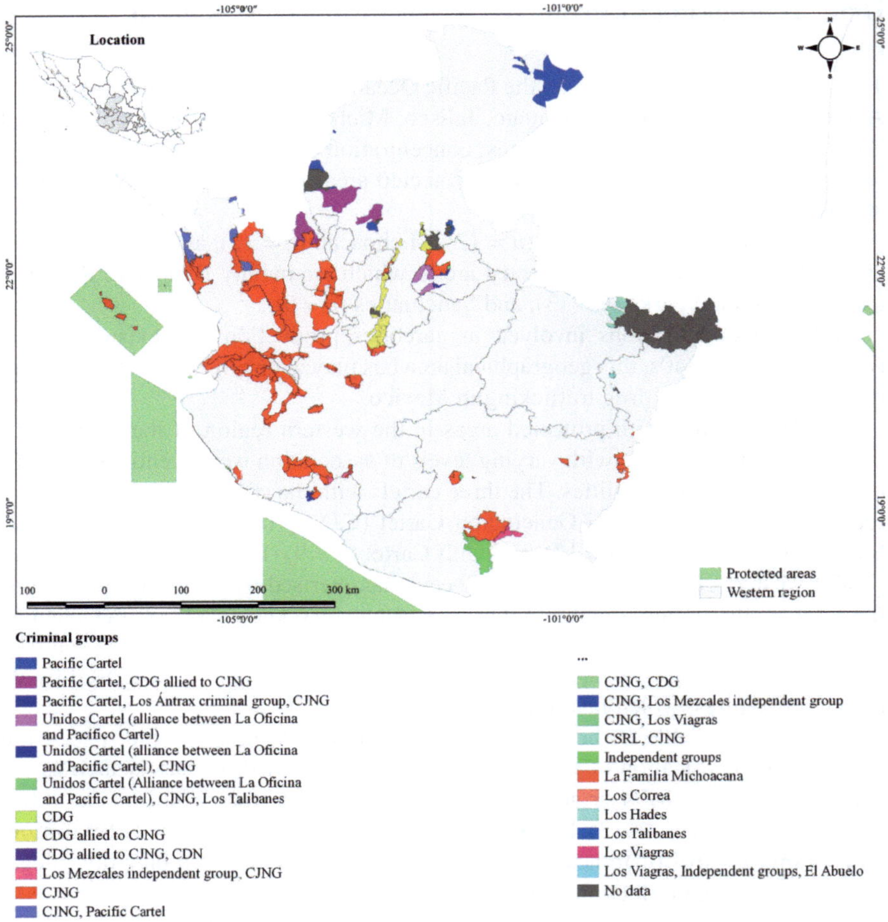

Fig. 9.6 Geographical distribution of drug cartels in protected areas in Western Mexico until 2024. **Source:** the authors with data from El Universal (2025) and CONANP (2025). **Notes:** a) The map shows only the municipalities where criminal groups converge in protected areas. The dominance of criminal groups certainly extends to other municipalities and states. b) The Sinaloa Cartel is also known as the Pacific Cartel

The emergence since 2009 of the Jalisco Nueva Generación Cartel (CJNG)[1] represented the main emerging criminal group with the highest growth and expansion. Under the leadership of Nemesio Oseguera Cervantes (alias "El Mencho"), the CJNG consolidated its position as the most violent and fastest-growing criminal group and represents the main enemy of other criminal groups that have not allied

[1] Initially, this criminal group was known as "Los Mata Zetas" (The Zetas Killers) and emerged as a group of direct and violent confrontation to eliminate the Zetas Cartel in this region.

with this criminal group (see Henkins, 2020; Jones, 2018; Niño & Pinto-Quijano, 2025; Sampó et al., 2023).

It should be highlighted that this region has been characterised by high levels of violence, the emergence of small criminal groups and self-defence groups, and pressure on rural and indigenous groups over territories for drug cultivation and trafficking, as well as other sources of economic income that are of interest to members of criminal groups. However, this region has also been notable for attacks on defenders of the environment and territories, mainly belonging to rural and indigenous communities.

According to data from CEMDA (2023b; 2024), states in this region concentrate attacks on conservationists and land defenders, mainly by organised crime groups. This is mainly due to the defence of forests and the corrupt association with government entities, which leads to the highest rates of deforestation in the country. Furthermore, there are no official records or information about the influence and impact of criminal organisations in the protected areas of this region.

Drug Trafficking in Protected Areas of the Central Region

The centre of the country is the region with the second highest number of criminal groups in municipalities with protected areas ($n = 146$ municipalities). In the central region, the State of Mexico ($n = 15$) and Mexico City ($n = 11$) are the states with the most protected areas, and the region is dominated by National Parks ($n = 34$) and Biosphere Reserves ($n = 8$).

In these municipalities with protected areas in the central region of the country, 40 different criminal groups with varying levels of association were identified. The three criminal groups with the strongest presence in the central region were: Jalisco Nueva Generación Cartel ($n = 28$ municipalities), La Familia Michoacana ($n = 22$), and Guerreros Unidos ($n = 7$) (Fig. 9.7).

This region includes the country's capital and its neighbouring states, favouring the concentration of criminal networks, connections, and groups involved in drug trafficking and other illegal activities. In this region is located the first protected area "Desierto de Los Leones" and it has forested and mountainous regions due to the Neovolcanic Axis in Central Mexico, but also deciduous forests and jungles near the coast in the Sierra Madre del Sur.

The Central region includes states historically linked to drug trafficking in Mexico. The area known as Tierra Caliente extends to the state of Guerrero, sharing this area with the state of Michoacán and the south of the State of Mexico. Furthermore, due to its location in the south-central Pacific region, it is considered a strategic area for the arrival of cocaine from Colombia by sea, as well as the precursor ingredients necessary for the creation of methamphetamines, amphetamines, and fentanyl.

The presence of criminal groups in this region (see Fig. 9.7), and particularly in protected areas, has been characterised by poppy production with the support of

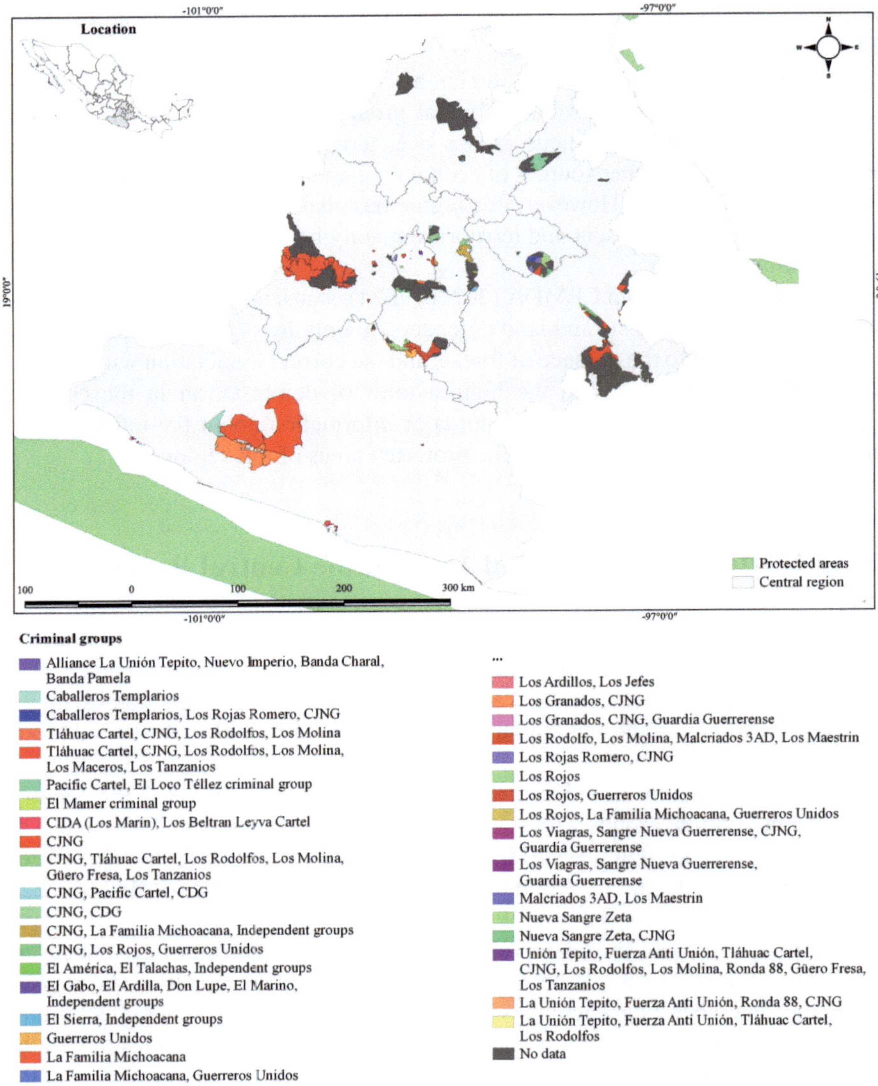

Fig. 9.7 Geographical distribution of drug cartels in protected areas in Central Mexico until 2024. **Source:** the authors with data from El Universal (2025) and CONANP (2025). **Notes:** a) The map shows only the municipalities where criminal groups converge in protected areas. The dominance of criminal groups certainly extends to other municipalities and states. b) The Sinaloa Cartel is also known as the Pacific Cartel

rural and indigenous communities, making the state of Guerrero the main poppy producer in the country, even surpassing the state of Sinaloa in the Golden Triangle (e.g. Gaussens, 2018; Saenz, 2024). In addition, this region is the second largest producer of marijuana nationwide and, together with poppy production, generates millions of dollars in profits and has promoted a discourse of social development in the indigenous and rural localities of the region due to the economic benefits generated by drug production.

Meanwhile, in the Mexico Valley region, the presence of criminal groups has expanded to other illegal activities besides drug trafficking (Instituto Nacional de Estadística y Geografía [INEGI], 2024). Criminal phenomena such as extortion, contract killings, hydrocarbon theft (huachicol), kidnapping, among others, have turned this region into one of the main crime hotspots in the country. Furthermore, these illegal activities take place in the region with the highest population density in the country and in natural areas that provide ecosystem services to cities and rural areas, promoting development and maintaining quality of life.

Drug Trafficking in Protected Areas of the Northwest Region

The northwestern region ($n = 109$) of Mexico is the third region with the most criminal groups in municipalities that have protected areas. In the northwestern region of Mexico, the states with the most protected areas are Chihuahua ($n = 14$) and Baja California ($n = 12$). In addition, most of the protected areas in this region are Flora and Fauna Protection Areas ($n = 22$) and Biosphere Reserves ($n = 15$).

In the municipalities with protected areas in this region, 15 different criminal groups with varying levels of association were identified, distributed across 109 municipalities with protected areas. The three cartels with the strongest presence were: Nuevo Cártel de Juárez, La Línea faction ($n = 27$ municipalities), Sinaloa (Pacific) Cartel ($n = 27$), and Sinaloa (Pacific) Cartel, Los Menores faction ($n = 25$) (Fig. 9.8).

This region has historically been the cradle of Mexican drug trafficking. However, it is the region that has changed the least in terms of the diversification of groups, which have instead fragmented into subgroups and armed groups dependent on the major cartels. In addition, the region's size and territorial extension confer access to the border with the United States and ecosystems and climates ideal for drug production.

The orographic and geographical characteristics of this region allowed for the emergence and consolidation of drug trafficking since its emergence. The Golden Triangle located in this region led the production of poppies and marijuana for more than 50 years and continues to be the main geographical region producing marijuana nationally and trafficking drugs to the state of California and the central region of the United States.

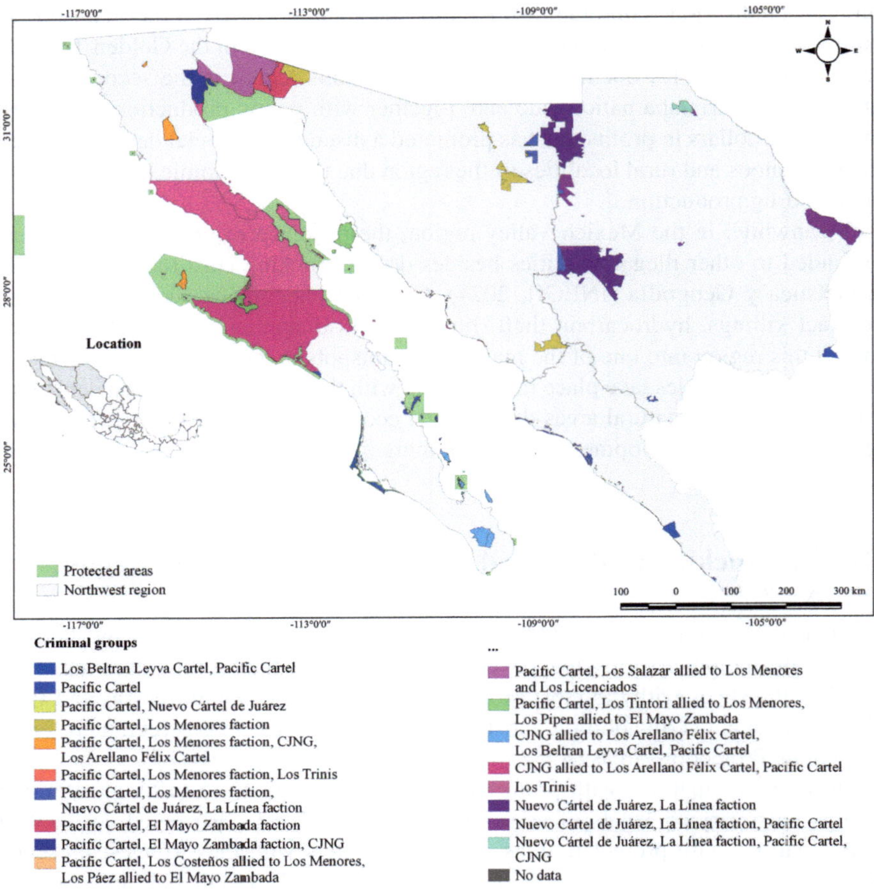

Fig. 9.8 Geographical distribution of drug cartels in protected areas in the Northwest region of Mexico until 2024. **Source:** the authors with data from El Universal (2025) and CONANP (2025). **Notes:** a) The map shows only the municipalities where criminal groups converge in protected areas. The dominance of criminal groups certainly extends to other municipalities and states. b) The Sinaloa Cartel is also known as the Pacific Cartel

The struggle to control strategic border territories for smuggling drugs into the United States, as well as remote wilderness areas (hidden) from urban centres, has led to violent battles for control and expansion among criminal groups. Some of the main conflicts between these criminal groups include the internal conflict within the Sinaloa Cartel (Pacifico) between "Los Chapitos" the leaders and sons of "El Chapo" Guzmán, and the Mayo Zambada faction known as "Los Mayos" or "La Mayiza" (see Mendoza-Cortés, 2020).

This conflict has led to one of the most recent outbreaks of violence in the region. Even the operations under Andrés Manuel López Obrador's policy of "hugs, not bullets" led to "Black Thursday" in October 2019, when a son of "El Chapo",

Ovidio Guzmán was arrested and released within a few hours due to pressure and violence exerted by the cartel on the federal and state governments, which ordered his release (see Mendoza-Cortés, 2020).

Although Ovidio Guzmán was arrested again in 2023 and Ismael "El Mayo" Zambada was arrested in 2024, both imprisoned in the United States, Ovidio Guzmán was extradited, while the Mexican government denied involvement in El Mayo's arrest, arguing that the federal government had not participated in the arrest and extradition (Villegas, 2025).

This region continues to be one with the most deeply rooted drug trafficking culture, not only as a business model, but also as a lifestyle in rural and indigenous communities, which depend on the economic benefits of drug trafficking due to historical neglect by state and federal governments. Protected areas continue to be strategic territories for drug production, but also as hideouts for members of criminal groups.

Drug Trafficking in Protected Areas of the Southeast Region

The Southeast region is the fourth region with the most criminal groups in municipalities that have protected areas (n = 105). In this region, Quintana Roo (n = 26) and Chiapas (n = 20) are the states with the most protected areas, which are mainly Flora and Fauna Protection Areas (n = 28), followed by Biosphere Reserves (n = 23) and National Parks (n = 23).

Southeastern Mexico is characterised by great biological and ecosystem diversity, with a mainly tropical climate that is home to a large concentration of biodiversity (e.g. Pérez-Irineo & Santos-Moreno, 2013; Schmitter-Soto et al., 2016). Although this region has been less associated with drug trafficking (at least in the collective consciousness), the presence of criminal groups in protected areas is just one of the many threats to environmental conservation.

Extractivism, mega-projects for "development" (both governmental and private), poverty, paramilitary groups, and the hydrocarbon industry (both legal and illegal), among others, represent challenges that are evidenced in various forms of illegal trafficking (wildlife, firearms, drugs, humans), homicides, and threats against environmental conservationists, deforestation, etc. All these phenomena originate in the private sector (industry, mining, etc.), the government and criminal groups (see CEMDA 2023b; 2024; Carpio-Domínguez & Castro-Salazar, 2025).

In municipalities with protected areas in the southeastern region of the country, only 11 different criminal groups were identified, with varying levels of association, but covering 105 municipalities in the states of Campeche, Quintana Roo, Chiapas, Oaxaca, Veracruz, Tabasco, and Yucatán. The three criminal groups with the strongest presence in the southeast region were: the Pacific Cartel (n = 21 municipalities), the Zetas Cartel 35-Z (n = 16), and the Jalisco New Generacion Cartel (n = 13) (Fig. 9.9). As in the other regions analysed, alliances between criminal groups were identified based on strategies for controlling territories, trafficking routes, or the

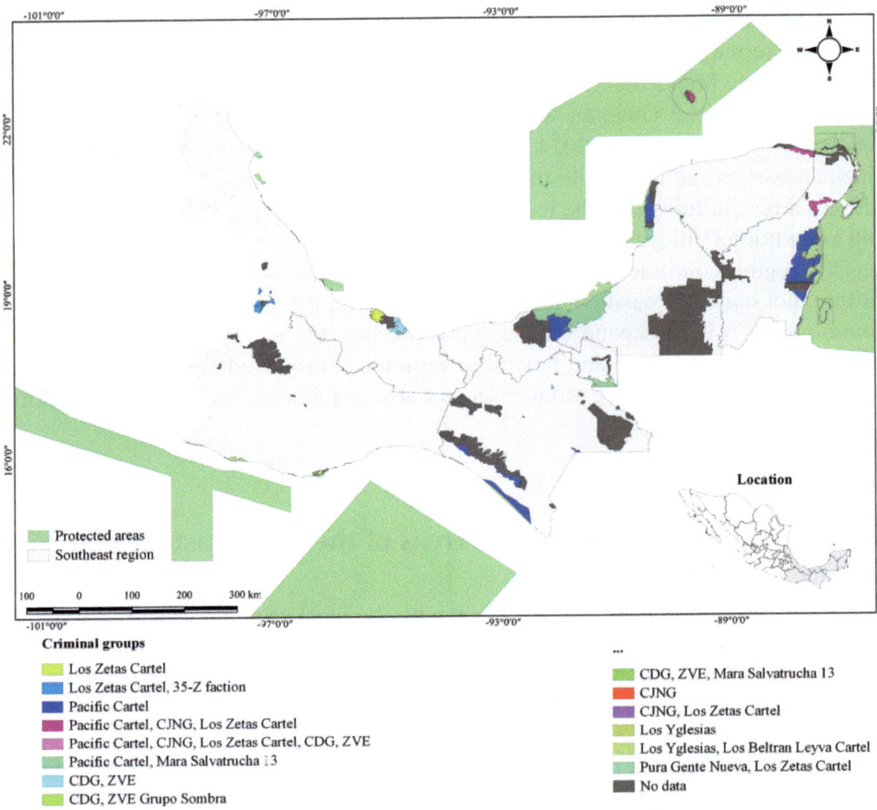

Fig. 9.9 Geographical distribution of drug cartels in protected areas in the Southeast region of Mexico until 2024. **Source:** the authors with data from El Universal (2025) and CONANP (2025). **Notes:** a) The map shows only the municipalities where criminal groups converge in protected areas. The dominance of criminal groups certainly extends to other municipalities and states. b) The Sinaloa Cartel is also known as the Pacific Cartel

struggle for legal and illegal products such as mining, logging, fishing, and extortion ("cobro de piso"), among others.

It was identified that criminal groups such as Los Mara Salvatrucha 13 are present in this region as a group associated with Mexican drug cartels (Gulf Cartel [CDG] and Los Zetas Vieja Escuela [ZVE]). This may be related to the fact that Mexican immigration policies are generally lax and have led not only to irregular migration, but also to Central American criminal groups settling in the country and finding a niche in Mexican drug trafficking to collaborate and operate in illegal activities (see Moreno, 2023; Roberts & Romaniuk, 2023).

Despite the region's climate and terrain, drug production is not the primary activity in the region. Historically, this region has been used as a transit route for drugs coming from Central America, although drug production in some areas cannot be discarded. This region has the country's main tourist areas (the Mexican Caribbean) and some population concentrations in medium-sized cities, but it is mainly wild territory with rural and indigenous communities.

Drug Trafficking in Protected Areas of the Northeast Region

The Northeast region ranks fifth in terms of the number of criminal groups operating in municipalities with protected areas. Eighteen different criminal groups with varying levels of association were identified in municipalities with protected areas in the Northeast region of the country, distributed across 100 municipalities with protected areas. The three cartels with the strongest presence in this region were: Gulf Cartel ($n = 21$ municipalities), Jalisco Nueva Generacion Cartel (CJNG) ($n = 13$), and Los Zetas Vieja Escuela (ZVE) ($n = 13$) (Fig. 9.10).

This region bordering the United States is the closest border from the centre of the country, making access to the border with the state of Texas faster and more accessible (in terms of time and distance) (e.g. Correa-Cabrera, 2016; Carpio-Domínguez et al., 2022, 2023, 2025). Although the Gulf Cartel was the last to emerge on the scene of the country's major drug cartels (alongside the Sinaloa Cartel, Juárez Cartel, and Tijuana Cartel), this cartel filled a strategic territorial void for drug trafficking due to the geographical location of this region. With access by sea in the Gulf of Mexico to receive drugs from Central America and its proximity to the state of Texas, criminal groups in this region supply drugs to the eastern United States.

After the fragmentation of the Gulf Cartel, Los Zetas, the Noreste Cartel, and other smaller groups have made this region one of the most dangerous in the country during the Mexican War on Drugs, and this situation continues to this day. In the last two presidential terms (see Fig. 9.2), hydrocarbon trafficking (huachicol) emerged as a new illicit market within these criminal groups in association with federal government officials, the Navy, and public servants at the state level (see Trejo-Mejía & Uribe-Pineda, 2024; Mexicanos contra la Corrupción, 2025).

In this region, Coahuila ($n = 9$) and San Luis Potosí ($n = 7$) are the states with the most protected areas, and the region mostly has Flora and Fauna Protection Areas ($n = 7$), Biosphere Reserves ($n = 7$), and National Parks ($n = 6$). The orographic and natural characteristics of this region are home to a large number of flora and fauna species and ecosystems and microhabitats of high value for environmental conservation. The presence of criminal groups in protected areas threatens the ability of environmental institutions such as CONANP to carry out environmental conservation actions. This particular area includes the state of Durango, historically dominated by the Sinaloa (Pacific) Cartel, and the border area with the state of Coahuila has been characterised by conflict between criminal groups for control of the territory.

Furthermore, the state with the highest representation in drug trafficking in the region is Tamaulipas. This state is the centre of a range of illegal activities such as drug trafficking, human trafficking (migrants), hydrocarbons ("huachicol"),

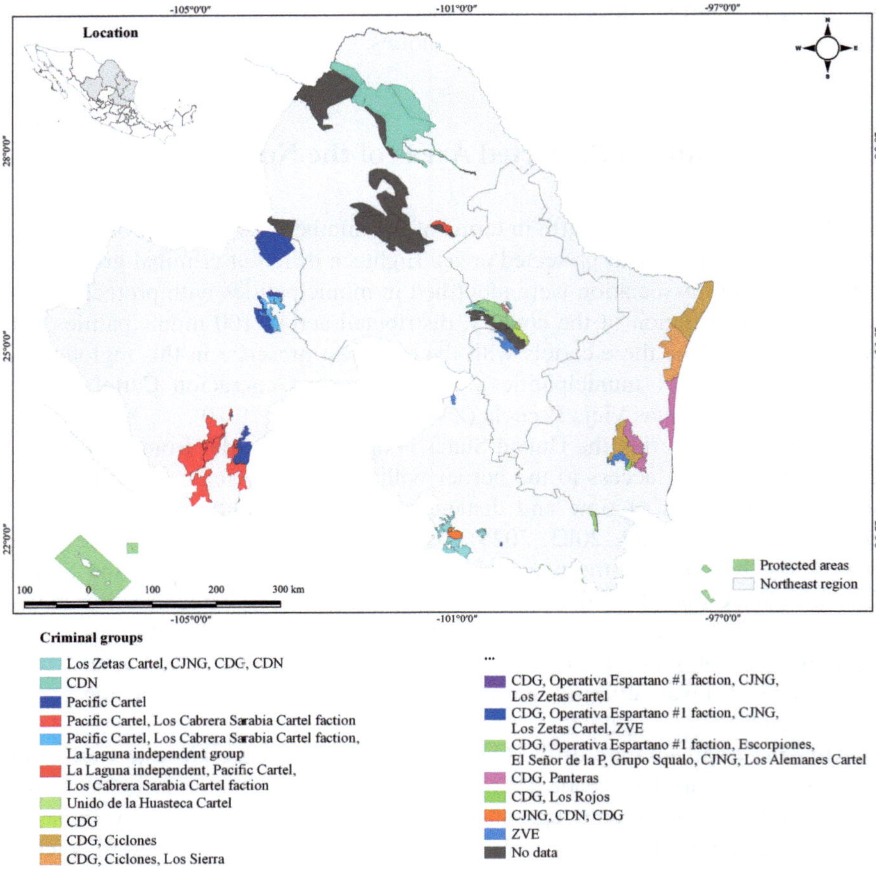

Fig. 9.10 Geographical distribution of drug cartels in protected areas in the Northeast region of Mexico until 2024. **Source:** the authors with data from El Universal (2025) and CONANP (2025). **Notes:** a) The map shows only the municipalities where criminal groups converge in protected areas. The dominance of criminal groups certainly extends to other municipalities and states. b) The Sinaloa Cartel is also known as the Pacific Cartel

firearms, and wildlife (e.g. Correa-Cabrera, 2016; Arroyo-Quiroz & Wyatt, 2019; Carpio-Domínguez et al., 2022, 2023, 2025; Morales de León et al., 2024). Furthermore, historical agreements between politicians and drug traffickers and its geographical location with access to the United States border and the Gulf of Mexico have made the state a coveted territory for criminal groups in Mexico.

However, the presence of criminal groups in protected areas in this region has not been documented in scientific studies or the media, making this study (see Fig. 9.10) the first formal mapping and study of the convergence of organised crime in the region's protected areas.

Distribution of Criminal Groups by Category of Protected Areas in Mexico

Considering the categories of protected areas in Mexico: Flora and Fauna Protection Areas (APFF its Spanish abbreviation), Natural Resource Protection Areas (APRN), Natural Monuments (MN), National Parks (PN), Biosphere Reserves (RB), and Sanctuaries (SANT) (Íñiguez Dávalos et al., 2014; CONABIO, 2023; Article 46 of the Ley General del Equilibrio Ecológico y Protección al Ambiente [LGEEPA], 2024) the following was identified:

In the Northwestern region, Flora and Fauna Protection Areas (APFF) have the highest number of criminal groups (n = 106 groups, 56.1%), while in the Central and Southeastern regions, National Parks (PN) have the highest number of criminal groups (n = 117, 56.8% and n = 59, 34.7%, respectively), in the West, Natural Resource Protection Areas have the highest number of criminal groups (n = 95, 43.8%), and in the Northeast, Biosphere Reserves (RB) have the highest number of criminal groups (n = 43, 26.5%) (Fig. 9.11).

In the southeast of the country, the Pacific Cartel is the criminal group with the largest presence in municipalities that have three of the six categories of protected areas (APFF, RB, and APRN). In the West, the Jalisco Nueva Generación Cartel has the largest presence in five of the six categories of protected areas (APFF, RB, PN, and SANT). In the Centre, the Jalisco Nueva Generación Cartel has the largest presence in three of the six categories of protected areas (RB, PN, and SANT). In the Northeast, the Gulf Cartel has the strongest presence in three of the six categories of protected areas (RB, PN, and MN). In the Northwest, the Sinaloa Cartel (Pacific) has the strongest presence in five of the six categories of protected areas (APFF, RB, PN, SANT, and MN).

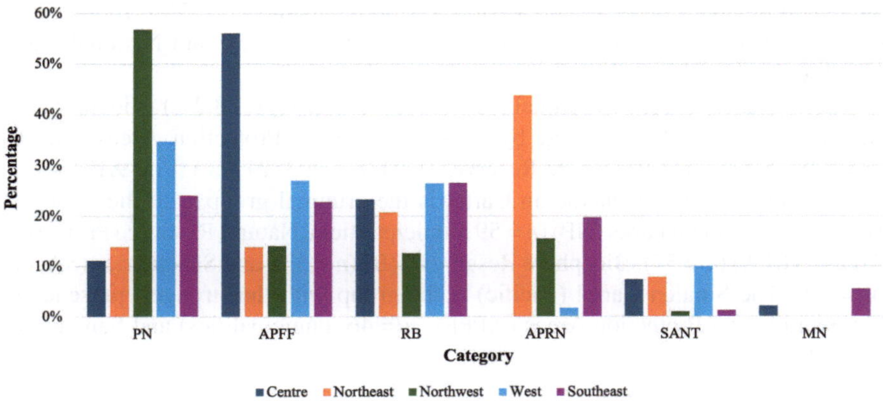

Fig. 9.11 Concentration of criminal groups in municipalities with protected areas, by category and region. **Source:** the authors with data from El Universal (2025) and CONANP (2025)

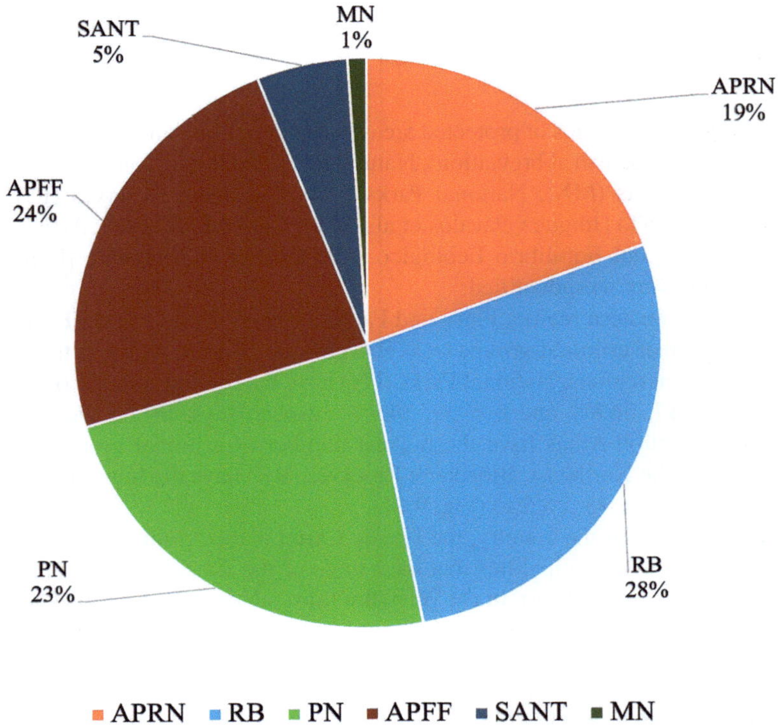

Fig. 9.12 Percentage of criminal groups in municipalities with protected areas by category in Mexico until 2024. Source: the authors with data from El Universal (2025) and CONANP (2025)

Biosphere Reserves (BR) have the highest presence of criminal groups within municipalities with protected areas (n = 160 municipalities, 27.5%), followed by Flora and Fauna Protection Areas (APFF) (n = 137, 23.6%) and National Parks (n = 136, 23.4%) (Fig. 9.12).

The highest number of criminal groups (n = 266 groups, 28.2%) is identified in National Parks (NP), followed by Flora and Fauna Protection Areas (APFF) (n = 248, 23.6%) and Biosphere Reserves (BR) (n = 203, 21.5%) (Fig. 9.13).

The Jalisco Nueva Generacion Cartel is the criminal group with the strongest presence in National Parks (NP) (n = 59 municipalities), Natural Resource Protection Areas (NRPA) (n = 57), Biosphere Reserves (RB) (n = 53), and Sanctuaries (SANT) (n = 10). The Sinaloa Cartel (Pacific) is the group with the strongest presence in Flora and Fauna Protection Areas (APFF) (n = 46 municipalities) and Sanctuaries (n = 10).

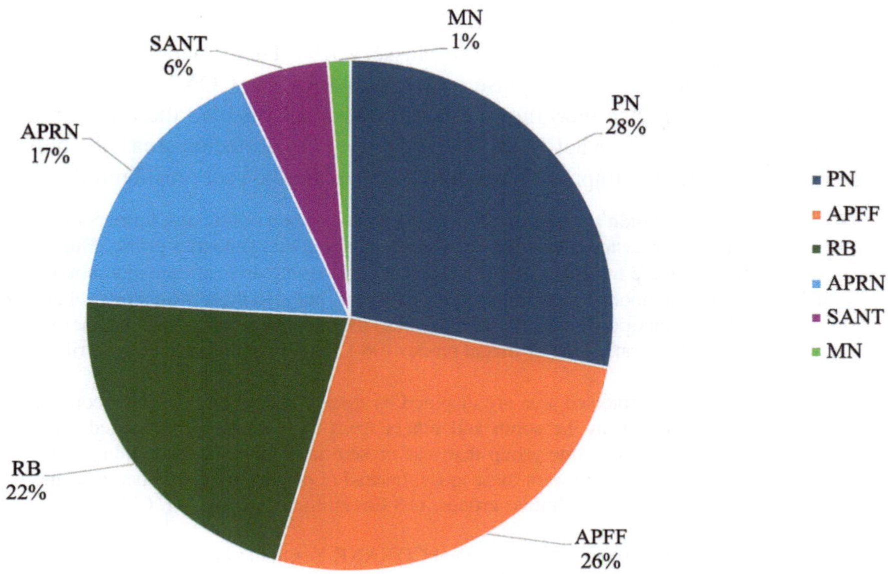

Fig. 9.13 Percentage of diversity of criminal groups in municipalities with protected areas, by category in Mexico until 2024. Source: the authors with data from El Universal (2025) and CONANP (2025)

Experiences of Environmental Conservation, Drug Trafficking in Mexico

Considering the current public security context in Mexico, the increase in criminal groups, the expansion of their territories, the emergence of new illegal trafficking, disappearances of persons, corruption, and budget cuts for environmental government institutions are analysed through the experiences of CONANP environmental inspectors and directors.

During the last presidential term (2018–2024), environmental conservation activities in Mexico's protected areas continued in a context of insecurity and vulnerability, with the particularity that environmental directors and inspectors already had 18 years of experience in environmental conservation in violent environments.

> In the ten years that I have been working in the protected area, I can tell you that nothing has changed. We continue to work under the same conditions of insecurity. The drug traffickers have always been here, they are still here, and they will continue to be here [...] the best we can do is our job (15-Environmental inspector, CJNG Cartel territory).
>
> Now it's worse. Before, [drug traffickers] only focused on drugs, but now they are avocado farmers. They have taken over territories legally through threats and forced displacement, causing wildfires to change land use, and the authorities pretend not to see this whole situation (12-Environmental inspector, territory of the La Familia Cartel).

Criminal groups currently view natural resources and territories as alternative sources of income to drug trafficking (see Massé, 2019). This situation has severely limited the implementation of conservation activities by CONANP personnel. Although their experience working in conflict zones has enabled them to conduct environmental conservation activities more effectively in protected areas, the public security situation has not improved over the course of the six-year presidential terms.

> Now we see things we didn't see before. Now drug traffickers are politicians, farmers (avocado and coffee), businessmen; they get involved in anything that generates profits. But let me tell you that there are two types of drug traffickers: those who drive around in armoured vehicles with hitmen, hooded men and military uniforms, but also those who work behind the scenes, who go unnoticed, who are businessmen, who engage in other legal businesses with money from drug trafficking or illegal money (08-Director of protected area, territory of the Sinaloa Cartel).
>
> Some parts of this protected area are disputed by three criminal groups, some coming from the north, others from the south and others from the east of the protected area. Previously, we knew of only one group that was present in some rural communities, but now there are three groups that want these areas. Instead of decreasing in number, it seems that there are more and more criminal groups. (16-Environmental inspector, CJNG Cartel territory).
>
> Honestly, it has been like this since I joined CONANP. Yes, it has become more dangerous recently, but I don't think things will change anytime soon. For me, this is not just a job, it's what I love to do, and I will continue doing it until I can no longer do so (15-Environmental inspector, CJNG Cartel territory).

Mexico's public security policy of "hugs, not bullets" has failed to reduce the social and environmental vulnerability of Mexico's protected areas. Environmental inspectors and directors do not have health insurance, only that provided by the State to public servants. Furthermore, this situation of vulnerability extends beyond personal safety; for environmental inspectors and directors, the safety of their families is a constant concern when conducting environmental conservation activities in protected areas dominated by organised crime.

> When I was a single woman, I was braver. Now I think about my children and ask God every day to let me return to see my children (02-Environmental inspector, Sinaloa Cartel territory).
>
> Not long ago, sicarios stopped us and pointed their guns at us. All I could think about was my wife and children. How would they get by without me? What if they never found my body, as happens to so many people in this country? Having a family changes your perspective on everything (16-Environmental inspector, CJNG Cartel territory).
>
> When I became a mother, I became more fearful. Not for myself, but for my children. Now that things are getting worse, I don't even want to think about one day being gone (dying) (09-Director of protected area, Gulf Cartel territory).
>
> Work in this protected area is not so dangerous, as we all know each other. We know the people in the communities, they recognise us, but they also recognise "la maña" [drug traffickers] and they know the people in the communities. We are not friends, but at least they do not bother us (13-Director of protected area, Sinaloa Cartel territory).

Public and personal insecurity caused by the presence of criminal groups in protected areas is evident in both dimensions of public security (Huerta-Benites, 2014). In the objective dimension of public safety, this is evident in physical attacks on environmental inspectors, but also in government actions such as budget cuts for environmental institutions (see Fig. 9.2) and the implementation of public policies that place inspectors in vulnerable situations. In the subjective dimension of public security, this is evident in the perceptions, feelings, and experiences of CONANP personnel regarding their personal safety as civilians and as public servants.

From a green criminology perspective, conservation, protection, study, and defence of the environment are considered vulnerable activities in Mexico (Carpio-Domínguez & Castro-Salazar, 2025; Gallegos-Martínez et al., 2023). Environmental inspectors and directors are a vulnerable population who are not only responsible for environmental conservation activities in the public sector but must also ensure their own safety through self-taught strategies inherited from their predecessors.

Environmental conservation activities within protected areas are affected by conditions of national public insecurity. Therefore, the presence of criminal groups in protected areas poses a threat to ecosystems, biodiversity, human communities (indigenous and rural), and government environmental inspectors.

> Recently, drug traffickers have found an economic niche in deforestation here in this protected area. It is the main problem in this area; they do not care about biodiversity or ecosystem services. They are logging trees and paying bribes to sell the timber (16-Environmental inspector, CJNG Cartel territory).
>
> We have found animal cadavers with gunshot wounds. We suspect that they are being hunted or shot for fun, I don't know (14-Environmental inspector, Los Zetas cartel territory).
>
> When we found a drug laboratory in the protected area, the Federal Attorney General's Office for Environmental Protection (PROFEPA) identified contamination from the chemicals used to process the drugs. But they also leave rubbish in the camps, throw away vegetable oil (which they use for cooking), drive in with their racers and crush the seedlings and cacti in the area, and they also like to wash their racers in the streams, so you can imagine the oil pollution in the water (15-Environmental inspector, CJNG Cartel territory).

Since the emergence of protected areas, drug trafficking has been present to this day. However, as evidenced by the experiences of environmental inspectors and directors, the conditions for implementing environmental conservation activities continue to be risky and have even increased due to the government's weakness in confronting organised crime in the country. The experiences recounted by environmental inspectors reveal that they have a more resilient view of the phenomenon of drug trafficking and organised crime in protected areas, but also a sense of resignation that change will not be achieved in the short term.

The experiences of the inspectors show that these professional activities for environmental conservation in Mexico's protected areas are a vocation that transcends many of the phenomena analysed in this book. The vocation for environmental conservation has ensured that environmental conservation activities are implemented even with institutional limitations such as neglect of the problems present in protected areas and insecurity caused by criminal groups. Even environmental inspectors and directors remained active in their roles throughout the Mexican War on Drugs and now during the period of "hugs, not bullets".

Conclusions

Public insecurity in Mexico continues to be present in all regions of the country. The heterogeneity of social, economic, political, cultural, geographical, and environmental contexts has meant that there is no single version of drug trafficking in nationally protected areas. Public policies on insecurity based on the slogan "hugs, not bullets" have not been effective in addressing the phenomenon of drug trafficking in Mexico but have instead led to its entrenchment even in natural environments such as protected areas.

The increase in criminal groups in the country highlights not only the need to address the problem at a social and governmental level, but also at an environmental level, as it puts Mexico's natural wealth at risk. The activities of criminal groups, initially limited to drug trafficking, have expanded into other legal and illegal activities that endanger and threaten environmental conservation in the country.

An analysis of drug trafficking and criminal groups in Mexico's protected areas during this period leads to the conclusion that criminal activities will find sources of income as long as there is demand for legal and illegal products, including drugs. The involvement of indigenous and rural communities in drug trafficking is a consequence of the government's neglect of basic needs, which have historically been focused on cities and semi-urban settlements, leaving rural and indigenous communities neglected in terms of basic needs and development.

The policy of "hugs, not bullets" has failed by being permissive towards criminal groups in the country. Although there is no single solution to the phenomenon of drug trafficking and organised crime in Mexico, it is evident that the government's neglect favours the presence, influence, and dominance of criminal groups in social life, governance, the economy and also in natural environments.

However, it is worth acknowledging and highlighting the resilience and hard work of CONANP's environmental inspectors and directors in conducting conservation activities in contexts of vulnerability, danger, and public and personal insecurity, as well as resisting the government's neglect of the institution's needs and the country's environmental problems.

References

Alvarado-Mendoza, A. (2024). La seguridad pública y la violencia durante la presidencia de Andrés Manuel López Obrador: Militarismo, criminalidad e imperio de la ley. *Cahiers des Amériques Latines, 104*. https://doi.org/10.4000/123ua

Arroyo-Quiroz, I., & Wyatt, T. (2019). Wildlife trafficking between the European Union and Mexico. *International Journal for Crime, Justice and Social Democracy, 8*(3), 23–37. https://doi.org/10.5204/ijcjsd.v8i3.1243

Barr, W. (2022). *One damn thing after another. Memoirs o fan Attorney General*. Harper Collins.

Buscaglia, E. (2015). *Lavado de dinero y corrupción política: el arte de la delincuencia organizada internacional*. Penguin Random House Grupo Editorial.

Cano, J. (2024). *Narcomapa 2024: qué cárteles tienen mayor presencia en México*. Infobae (23 de abril de 2024). México. https://www.infobae.com/mexico/2024/04/23/narcomapa-2024-que-carteles-tienen-mayor-presencia-en-mexico/
Carpio-Domínguez, J., & Castro-Salazar, J. (2025). Researching environmental harms in violent environments of the global south. In R. P. Cavalcanti, D. S. Fonseca, V. Vegh Weis, K. Carrington, R. Hogg, & J. Scott (Eds.), *The Palgrave handbook of criminology and the global south*. Palgrave Macmillan. https://doi.org/10.1007/978-3-031-74932-2_115-1
Carpio-Domínguez, J., Arroyo-Quiroz, I., Villarreal-Martínez, M. y Castro-Salazar, I. (2022). Tigers of Mexican drug traffickers, national statistical review and a case study through network analysis of the possession of a tiger (*Panthera tigris*) in northeastern Mexico. Forensic Science International: Animals and Environments, 2, 100039. https://doi.org/10.1016/j.fsiae.2021.100039.
Carpio-Domínguez, J., Cervantes-Niño, J., Castro-Salazar, J., & Mendezcarlo-Silva, V. (2025). Policing wildlife trafficking in northeastern Mexico: The case of Tamaulipas in 2023-2024. *Frontiers in Conservation Sciences, 6*(2025), 1488500. https://doi.org/10.3389/fcosc.2025.1488500
Carpio-Domínguez, J., Vargas-Orozco, C., & Villarreal-Sotelo, K. (2023). Redes de tráfico ilegal de vida silvestre en el noreste mexicano, un análisis desde la criminología verde. *Revista Española de Investigación Criminológica, 21*(2), e789. https://doi.org/10.46381/reic.v21i2.789
Centro Mexicano de Derecho Ambiental (CEMDA). (2023a). *Más áreas naturales protegidas y menos recursos públicos para su conservación*. CEMDA. https://cemda.org.mx/mas-areas-naturales-protegidas-y-menos-recursos-publicos-para-su-conservacion/
Centro Mexicano de Derecho Ambiental (CEMDA). (2023b). *Informe sobre la situación de las personas y comunidades defensoras de los derechos humanos ambientales en México 2023*. CEMDA. https://cemda.org.mx/informe-sobre-la-situacion-de-las-personas-y-comunidades-defensoras-de-los-derechos-humanos-ambientales-en-mexico-2023/
Centro Mexicano de Derecho Ambiental (CEMDA). (2024). *Informe sobre la situación de las personas y comunidades defensoras de los derechos humanos ambientales en México 2024*. CEMDA. https://cemda.org.mx/wp-content/uploads/2025/04/INFORMEDEFENSORES_2024_WEB.pdf
Comisión Nacional de Áreas Naturales Protegidas (CONANP). (2025). Sistema de Información, Monitoreo y Evaluación para la Conservación (SIMEC). Gobierno de México. Available on https://simec.conanp.gob.mx/.
Comisión Nacional de Búsqueda. (2025). *Versión estadística del Registro Nacional de Personas Desaparecidas y No Localizadas (RNPDNO) septiembre de 2025*. Comisión Nacional de Búsqueda. https://versionpublicarnpdno.segob.gob.mx/Dashboard/ContextoGeneral
Comisión Nacional para el Conocimiento y Uso de la Biodiversidad (CONABIO). (2023). Áreas protegidas. Available on https://www.biodiversidad.gob.mx/region/areasprot.
Confederación Patronal de la República Mexicana (COPARMEX). (2024). ¿Qué perdemos con la desaparición de los organismos autónomos? Derechos, transparencia y progreso. COPARMEX. Available on: https://coparmex.org.mx/que-perdemos-con-la-desaparicion-de-los-organismos-autonomos-derechos-transparencia-y-prograso/.
Correa-Cabrera, G. (2016). Militarización y violencia en Tamaulipas. *Atlas de la Seguridad y la Defensa de México*, 181–189. Available on https://scholarworks.utrgv.edu/pass_fac/15/
Cumplido-Tercero, M. (2015). *Evolución del crimen organizado en México. Más allá del narcotráfico*. Instituto Español de Estudios Estratégicos. Available on https://dialnet.unirioja.es/servlet/articulo?codigo=7687260
El Universal. (2025). Mapa interactivo del crimen organizado. Availabe on https://interactivos.eluniversal.com.mx/2023/mapa-crimen-organizado/.
Fuentes-Díaz, A. (2015). Narcotráfico y autodefensa comunitaria en "Tierra Caliente", Michoacán, México. *CienciaUAT, 10*(1), 68–82. http://www.scielo.org.mx/scielo.php?script=sci_arttext&pid=S2007-78582015000200068&lng=es&tlng=es

Gallegos-Martínez, G., Carpio-Domínguez, J. and Castro-Salazar, I. (2023). Environmental exploitation and violence against indigenous people in Mexico. In Goyes, D. (Ed.) Green crime in the global south. Palgrave studies in green criminology. Palgrave Macmillan, . https://doi.org/10.1007/978-3-031-27754-2_4.

Gaussens, P. (2018). La otra montaña roja: el cultivo de la amapola en Guerrero. *Textual: análisis del medio rural latinoamericano, 71*, 33–69. https://doi.org/10.5154/r.textual.2017.71.003

Gobierno de México. (2018). *Estrategia Nacional de Seguridad Publica*. Gobierno de México. Available on: https://infosen.senado.gob.mx/sgsp/gaceta/64/1/2019-02-01-1/assets/documentos/Estrategia_Seguridad.pdf.

Grillo, I. (2011). *El Narco: inside Mexico's criminal insurgence*. Bloomsbury Press.

Henkin, S. (2020). The violent rise of Cártel de Jalisco Nueva Generación (CJNG). National Consortium for the Studt of Terrorism and Responses to Terrorism. University of Maryland. Available on https://digitalcommons.fiu.edu/srhreports/toc/toc/120/.

Hernández-Morales, A. (2023). Disputa por la soberanía. Crimen Organizado y defensa comunitaria como gobiernos privados indirectos en México. *Acta Sociologica*, 88–89., 187–214. https://doi.org/10.22201/fcpys.24484938e.2022.88-89.84875

Hernández-Soc, A. (2019). De Tierra Caliente a la Sierra y Costa Chica de Guerrero: desplazamiento interno forzado. *Cultura y representaciones sociales, 14*(27), 143–182. https://doi.org/10.28965/2019-27-05

Hernández, D. (2024). Lamenta UNAM decreto de AMLO; perjudicial quitar facultades a CONABIO, alertan. La Silla Rota (16 de marzo de 2024). Available on: https://lasillarota.com/nacion/2024/3/16/lamenta-unam-decreto-de-amlo-perjudicial-quitar-facultades-conabio-alertan-474272.html.

Huerta-Benites, F. (2014). *La seguridad ciudadana. Una mirada integral*. Instituto de Economía y Empresa. Available on https://www.iee.edu.pe/doc/publicaciones/TemasDeDesarrollo/19-2014-03-SeguridadCiudadanaPNUD-IEE.pdf#:~:text=La%20dimensi%C3%B3n%20objetiva%20de%20la%20seguridad%20ciudadana,en%20el%20desarrollo%20humano%20de%20las%20personas

Instituto Nacional de Estadística y Geografía (INEGI). (2024). Censo Nacional de Seguridad Pública 2024. Gobierno de México. https://www.inegi.org.mx/contenidos/programas/cnspf/2024/doc/cnspf_2024_resultados.pdf.

Íñiguez Dávalos, L. I., Jiménez Sierra, C. L., Sosa Ramírez, J., & Ortega-Rubio, A. (2014). Categorías de las áreas naturales protegidas en México y una propuesta para la evaluación de su efectividad. *Investigación y Ciencia, 22*(60), 65–70. https://www.redalyc.org/pdf/674/67431160008.pdf

Jones, N. (2018). The strategic implications of the Cártel de Jalisco Nueva Generación. *Journal of Strategic Security, 11*(1), 19–42. http://www.jstor.org/stable/26466904

Jones, N., & Alonso-Ramos, J. (2025). The fentanyl disruption. Guerrero, organized crime, and opium. In J. Rosen & S. Cutrona (Eds.), *Fentanyl. From the labs to the streets*. Routledge. https://www.taylorfrancis.com/chapters/edit/10.4324/9781003631873-10/fentanyl-disruption-nathan-jones-josé-jonathan-alonso-ramos

Ley de la Guardia Nacional. (2025). Ultima reforma publicada en el Diario Oficial de la Federación el 16 de julio de 2025. Cámara de Diputados del H. Congreso de la Unión. Gobierno de México. https://www.diputados.gob.mx/LeyesBiblio/pdf/LGN.pdf.

Ley General del Equilibrio Ecológico y Protección al Ambiente (LGEEPA). (2024). Última Reforma Publicada en el Diario Oficial de la Federación en abril 4 de 2024. Congreso de la Unión. https://www.ordenjuridico.gob.mx/leyes.php#gsc.tab=0.

López-León, A. (2017). Rural civil society in a Mexican narco-violent border: Some initial reflections. *Acta Sociologica, 74*, 57–80. https://doi.org/10.1016/j.acso.2017.11.004

Maldonado-Aranda, S. (2012). Drogas, violencia y militarización en el México rural. El caso de Michoacán. *Revista Mexicana De Sociología, 74*(1). https://doi.org/10.22201/iis.01882503p.2012.1.29532

Maldonado-Aranda, S. (2014). Despejando caminos inseguros. Itinerarios de una investigación sobre la violencia en México. *Alteridades, 24*(47), 63–76. https://alteridades.izt.uam.mx/index.php/Alte/article/view/779

Massé, F. (2019). Conservation law enforcement: Policing protected areas. *Annals of the American Association of Geographers, 110*(3), 758–773. https://doi.org/10.1080/24694452.2019.1630249

Mendoza-Cortés, P. (2020). Inteligencia y contrainteligencia militar frente a fallos y desafíos. El caso de Culiacán, México (2019). *URVIO. Revista Latinoamericana de Estudios de Seguridad, 26*, 37–56. https://revistas.flacsoandes.edu.ec/urvio/article/download/4225/3260?inline=1

Mendoza, I. (2024). Narco Service Provisions in Tierra Caliente: La Familia Michoacana, Los Caballeros Templarios, and Los Zetas. *Res Publica-Journal of Undergraduate Research, 29*, 7. https://digitalcommons.iwu.edu/respublica/vol29/iss1/7

Mexicanos contra la Corrupción. (2025). *Huachicol fiscal*. Anuario de la Corrupción en Mexico. Available on https://contralacorrupcion.mx/anuario-de-la-corrupcion-2025-gobierno-de-sheinbaum/huachicol-fiscal-corrupcion-mexico/

México Evalúa. (2025). *Gasto público 2018–2024: no se cumplió la austeridad republicana*. Mexico Evalua: en analisis, anuales, finanzas públicas, sexenales. Available on: https://numerosdeerario.mexicoevalua.org/2025/02/11/gasto-publico-con-amlo-no-cumplio-su-austeridad-republicana/.

Montalvo, T. (2013). *El gobierno federal delimita las regiones de la estrategia de seguridad*. Periódico Expansión (10 de enero de 2013). https://expansion.mx/nacional/2013/01/10/el-gobierno-federal-delimita-las-regiones-de-la-estrategia-de-seguridad.

Morales de León, M., Vargas Orozco, C., & Villareal Sotelo, K. (2024). Migración climática en Latinoamérica. La frontera noreste de México como lugar de tránsito y destino. *Sociedad y Ambiente, 2024*(27), 1–33. https://doi.org/10.31840/sya.v2024i27.2874

Morales-Gámez, L., & Ortega-Ramírez, A. (2024). Security and interdependence between Mexico and the United States (2019-2023): How do they avoid declaring drug traffickers as terrorists? *Norteamerica, 19*(2), 219–240. https://doi.org/10.22201/cisan.24487228e.2024.2.656

Moreno, I. (2023). La Mara Salvatrucha presentada en el teatro mexicano contemporáneo. *reCHERches, 31*, 29–45. https://doi.org/10.4000/cher.15610

National Vital Statistics System. (2025). Provisional drug overdose death counts. United States. Availabe on https://www.cdc.gov/nchs/nvss/vsrr/drug-overdose-data.htm.

Niño, C., & Pinto-Quijano, Á. (2025). Integración regional criminal: aproximación a la relación entre el Cartel de Jalisco Nueva Generación y el Clan del Golfo. *Revista Criminalidad, 67*(1), 79–95. https://doi.org/10.47741/17943108.640

Pérez-Irineo, G., & Santos-Moreno, A. (2013). Riqueza de especies y gremios tróficos de mamíferos carnívoros en una selva alta del sureste de México. *Therya, 4*(3), 551–564. https://doi.org/10.12933/therya-13-157

Roa, J. (2020). Presencia de grupos criminales en México (2018-2019). *Programa de Política de Drogas.*. https://www.politicadedrogas.org/PPD/index.php/observatorio/metaDescargables/id/137.html

Roberts, R., & Romaniuk, S. (2023). Mara Salvatrucha (MS-13) international criminal gang. In S. Ramaniuk, S. Catino, & A. Martin (Eds.), *The handbook of homeland security*. Taylor and Francis-CRC Press. https://doi.org/10.4324/9781315144511

Saenz, A. (2024). *Guerrero: la montaña de amapola.* Pie de Pagina (February 21, 2024). Available on https://piedepagina.mx/guerrero-la-amapola-y-el-agua/.

Sampó, C., Jenne, N., & Ferreira, M. (2023). Ruling violently: The exercise of criminal governance by the Mexican Cartel Jalisco Nueva Generación (CJNG). *Revista Científica General José María Córdova, 21*(43), 647–665. https://doi.org/10.21830/19006586.1172

Schmitter-Soto, J., Mariaca-Méndez, R., & Soto-Pinto, M. (2016). Una breve historia del conocimiento y uso de la biodiversidaden la frontera sur de México. *Sociedad y Ambiente, 11*, 160–173. https://www.redalyc.org/pdf/4557/455748464009.pdf

Secretaría de Gobernación. (2013). Programa Sectorial de Gobernación 2013–2018. Gobierno de México. https://www.dof.gob.mx/nota_detalle.php?codigo=5326204&fecha=12/12/2013#gsc.tab=0.

Secretaría de Hacienda y Crédito Público (SHCP). (2025). Transparencia presupuestaria. Observatorio del gobierno. Gobierno de Mexico: SHCP. Available on https://www.transparenciapresupuestaria.gob.mx/.

Secretaria de Seguridad y Protección Ciudadana. (2018). Estrategia Nacional de Paz y Seguridad 2018–2024. Garantizar respeto y promoción de los Derechos Humanos. Gobierno de México. https://www.gob.mx/sspc/articulos/estrategia-nacional-de-paz-y-seguridad-2018-2024?idiom=es.

Serrano-Carreto, M. (2019). La estrategia de seguridad de AMLO. ¿De la pacificación a la militarización? *IUS Revista del Instituto de Ciencias Jurídicas de Puebla, 13*(44), 207–227. https://doi.org/10.35487/rius.v13i44.2019.615

Tovar, A. (2021). Dividen al país en 5 regiones para atender zonas violentas. Tribuna de la Bahía (21 de julio de 2021). https://tribunadelabahia.com.mx/dividen-al-pais-en-5-regiones-47728/.

Trejo-Mejía, K., & Uribe-Pineda, C. (2024). La extracción y venta de huachicol como trabajo ilícito: Un relato social de transformación regional (Tula de Allende, Hidalgo). *Psicología, Educación & Sociedad, 3*(6), 1–18. https://doi.org/10.5281/zenodo.14098566

Valdés, G. (2013). *Historia del narcotráfico en México*. Editorial Aguilar.

Vázquez Valdez, J. (2018). Economía del narco: prohibicionismo, violencias sistémicas y capital criminal. *Caleidoscopio-Revista Semestral De Ciencias Sociales Y Humanidades, 22*(38), 105–130. https://doi.org/10.33064/38crscsh917

Villegas, P. (2025). *La guerra interna del Cártel de Sinaloa, explicada*. New York Times (June 30, 2025). Available on https://www.nytimes.com/es/2025/06/30/espanol/america-latina/cartel-sinaloa-mexico-guerra-interna.html

Chapter 10
New Illicit Markets and Crimes in Natural Protected Areas

Introduction

In 2025, criminal groups continue to be the main cause of public insecurity in the country. At the end of Andrés Manuel López Obrador's six-year presidential term with the slogan "hugs, not bullets", Claudia Sheinbaum Pardo's presidential term continued with the same policy of non-confrontation with organised crime from the presidency of the republic. After seven years, the increase in criminal groups in the country to 175 (Roa, 2020; Cano, 2024) has led to the territorial expansion and consolidation of criminal groups such as the Jalisco Nueva Generacion Cartel and their incursion into other profitable legal and illegal activities. However, it should be emphasised that drug trafficking continues to be the most profitable illegal business in Mexico and the world.

With Donald Trump as president of the United States, political tensions with the Mexican government continued due to permissiveness and lack of direct action against drug trafficking. Furthermore, with fentanyl entering the Mexican drug trafficking scenario and having negative impacts on both countries (public safety in Mexico and health in the United States), the United States government continues to consider Mexican cartels as Foreign Terrorist Organizations (FTOs), primarily the Jalisco Nueva Generacion Cartel and the Sinaloa Cartel (Morales-Gámez & Ortega-Ramírez, 2024).

Drug trafficking and criminal groups in Mexico are characterised by their dynamism, constantly adapting to new opportunities to generate financial resources for their survival and operation. The "Mexican Variant"[1] of organised crime (Valdés,

[1] The Mexican Variant consists of criminal organisations that operate without the inconvenience of having the authorities against them, through a pact with the State, which only pursues them sporadically and selectively and allows them to operate freely without hiding, travel without problems on the country's roads, integrate vertically and horizontally into society and government institutions, and put police and other public security agents at their service.

J. L. Carpio-Domínguez et al., *Drug Trafficking and Protected Areas in Mexico*,
https://doi.org/10.1007/978-3-032-18218-0_10

2013, p. 173) (see Chap. 6) returned to Mexico in 2018 and has allowed for a resurgence of corruption at the government level, state neglect of public security issues, and the weakening of government institutions.

Although it cannot be discarded that there was an "adaptation" of the Mexican Variant of organised crime during the six-year terms of Felipe de Jesús Calderón Hinojosa (2006–2012) and Enrique Peña Nieto (2012–2018), it is relevant to this study to highlight that the policies of non-confrontation with drug trafficking under the slogan "hugs, not bullets" initiated in 2018 has allowed criminal groups to strengthen nationally and internationally, even incorporating practices of corruption and nepotism, weakening government institutions, budget cuts, and political polarisation of Mexican society (see Sarsfield, 2023; Moreno, 2024; Rivera-Magos & González-Pureco, 2024).

However, never in the history of the country and its protected areas has the impact of criminal groups and drug trafficking in Mexico's protected areas been formally documented (although partially addressed in Carpio-Domínguez, 2021). This chapter explores some of the illegal activities that have emerged in recent years as new illicit markets within and near protected areas, not only to formally document these illegal activities in the country, but also to understand the importance of protected areas as strategic territories for illegal activities by criminal groups and drug trafficking in Mexico.

Through documentary analysis, this chapter examines some of the main emerging illicit markets and crimes committed by criminal and drug trafficking groups in the country's protected areas. It also brings to light the experiences of environmental inspectors and directors from the Comisión Nacional de Áreas Naturales Protegidas (CONANP) (National Commission of Protected Natural Areas) during 2024–2025, allowing for a comparison of how environmental conservation practices have evolved in protected areas in a context of violence caused by drug trafficking and criminal groups.

Deforestation, an Old and New Issue in Mexico

Deforestation is the act of reducing or eliminating tree cover below the threshold value for forest and the human-induced change of land use from forest to some other nonforest use, and deforestation also occurs in ecosystems such as grasslands, deserts, mangroves, etc. (Carpio-Domínguez, 2024; Food and Agriculture Organization of the United Nations [FAO], 2010).

This practice affects and endangers ecosystems and biodiversity, leading to ecological disasters directly related to the reduction of wildlife, the alteration of biogeochemical cycles, including the ability to absorb and store nearly half of the world's carbon, and diminishes the ability to mitigate climate change (Van Solinge, 2010a, b, 2014).

Deforestation is linked to timber extraction, agricultural expansion, human settlements, urban growth, private and government infrastructure projects, and subsistence farming practices (España-Boquera & Champo-Jiménez, 2016). Although it is not only a local problem, but also a regional issue, deforestation has been documented in Latin America (e.g. Cotrina-Sánchez et al., 2021; Mujica et al., 2024) and other parts of the world (Balboni et al., 2023; Cheeseman, 2024; Santos-Duarte et al., 2021).

Techniques for deforestation include dragging chains, logging with tools such as chainsaws, hand saws, etc., but one of the most common techniques is through wildfires and burning the wild environment. In Mexico, intentional forest wildfires are one of the main causes of land use change, with the aim of changing land use from forest to nonforest for agriculture (seasonal and intensive) and urban expansion of new neighbourhoods and housing developments due to population growth in urban centres.

Between 2010 and 2024, 45,560 fires were reported in Mexico, of which 20.3% (n = 9268) occurred in protected areas (Fig. 10.1) (Comisión Nacional Forestal (CONAFOR) (2024) (National Forestry Commission). 82.4% (n = 7644) of wildfires within protected areas were caused by human activities and 11.8% (n = 1101) by unknown/undetermined causes. Among the wildfires caused by humans, those caused by agricultural activities (33.1%, n = 3076), intentional acts (19.5%, n = 1810), and illegal activities (8.3%, n = 770) can be highlighted.

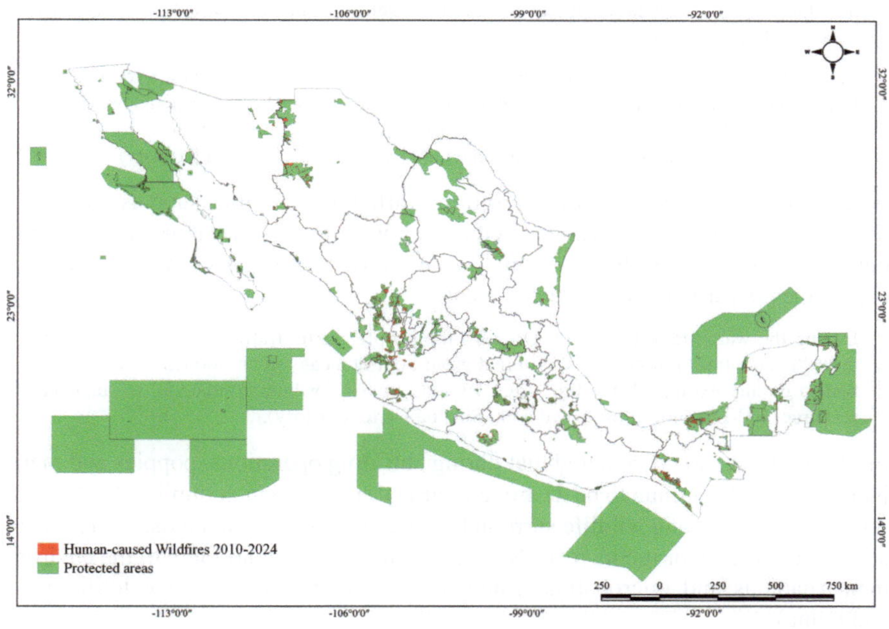

Fig. 10.1 Forest wildfires in protected areas in Mexico during 2010–2024. Source: the authors with data from Comisión Nacional Forestal (CONAFOR) (National Forestry Commission) (2024) and CONANP (2025)

The two states that have had the most unintentional or negligent forest wildfires in protected areas are the Estado de México (n = 1658) and Chiapas (n = 1226), while the two states with the most intentionally caused wildfires within protected areas are the Estado de México (n = 410) and Jalisco (n = 334). However, it is possible that accidental wildfires were actually intentional due to a lack of surveillance and specialised analysis to identify the causes of the wildfires (Maguey & Francisco, 2022).

According to data reported by the Comisión Nacional Forestal (CONAFOR) (2024) (National Forestry Commission), the two states with the most wildfires within protected areas caused by illegal activities are the Estado de México (n = 267) and Morelos (n = 145). Due to methodological limitations, criminal investigation, and criminal law enforcement related to wildfires, Mexico has prioritised the creation of firefighting and immediate response brigades. Protected areas have firefighting brigades that have improved over time and across administrations, but they continue to face serious limitations in terms of access to equipment and tools to fight wildfires.

> Budget cuts have hit us hard. I can't afford to equip my boys [community volunteers] [...] There's no way I can send them out to fight wildfires wearing huaraches [sandals]. They do more than enough just by supporting what we do here. (02-Environmental inspector, Sinaloa Cartel territory).

> Something that has affected us here in the protected area is intentional wildfires. In fact, I had to pay for training out of my own pocket to learn everything about forest wildfires [...] now there are no financial resources for training public servants (04-Director of protected area, Sinaloa Cartel territory).

> The best we can do is extinguish them [forest wildfires]. We do not know how to conduct or initiate an investigation to identify the perpetrators, nor have we been instructed on how to do so. Although we are aware of who is responsible [drug traffickers], without evidence, we are unable to prove anything (15-Environmental inspector, CJNG Cartel territory).

Forest fires were previously linked to drug trafficking activities such as local conflicts over territory, which were started as a strategy to expel rival groups from disputed territories. CONANP personnel made efforts to extinguish the wildfires and ensure the conservation of protected areas.

> During the war [Mexican War on Drugs], "the people" [drug traffickers] set wildfires to drive "the contras" [another rival criminal group] out of a canyon or wooded area. And we had to go and extinguish the fire, often without security or human, material or financial resources (03-Director of protected area, Sinaloa Cartel territory).

Furthermore, it should be noted that during anti-drug operations, poppies and marijuana were and continue to be destroyed (burned) in situ, which implies the death of pollinating insects and wildlife surrounding the drug cultivation fields. There are no academic, official, or media records documenting the harm not only to the environment (habitats and microhabitats) and human safety, but also to wildlife (flora and fauna).

Currently, forest fires are also linked to other illegal activities such as sawmills for timber production and the incursion of criminal groups into the avocado, coffee,

Fig. 10.2 Mountain ecosystems in protected areas of northwestern Mexico. Photographs: José Luis Carpio Domínguez

and citrus fruit (such as lemons and oranges) sectors, with all the environmental harm that deforestation entails: removal of vegetation cover, alteration of soil properties, loss of native flora, loss of fauna, mycorrhizae, fungi, and microfauna that maintain the environmental health of ecosystems (Fig. 10.2).

(Narco)terrorism and Agricultural Monopolies

With the diversification of criminal groups' activities in Mexico (Buscaglia, 2015; Cumplido-Tercero, 2015; Fuentes-Díaz, 2015; Vázquez-Valdéz, 2018), due to the continued failure of national and international government policies to eradicate drug trafficking at the production level rather than at the demand level (see Cameron, 2019; McSweeney et al., 2014) and the lack of confrontation exacerbated by the policy of "hugs, not bullets" of the last two presidential terms (Sarsfield, 2023; Moreno, 2024; Rivera-Magos & González-Pureco, 2024), organised crime has had free rein to enter all possible profitable markets in Mexico.

With its sights set on agricultural supply chains, organised crime has taken over this sector in Mexico, controlling the production, distribution, and prices of avocados and, more recently, citrus fruits such as lemons and oranges, as well as coffee. Criminal groups such as the Jalisco Nueva Generación Cartel, La Familia (Michoacana) Cartel, and Los Caballeros Templarios exercise narco-terrorist control over agricultural communities (Carbajal, 2023).

Agricultural communities producing avocados, lemons, and coffee in states such as Michoacán, Guerrero, Oaxaca, Veracruz, and Chiapas are victims of illegal and narco-terrorist activities by criminal groups, such as extortion, through which they force producers and packers to pay a fee for every kilo of lemon or avocado harvested or sold (see Váldes & Sánchez, 2020). Criminal groups also exercise control over the supply chain, imposing schedules for harvesting avocados or lemons and controlling transport routes. Those who oppose this control are disappeared or murdered to prevent uprisings by people in self-defence and to instil terror and dominate communities and the agricultural business (Fig. 10.3).

Furthermore, as with avocado and lemon production, criminal groups illegally and forcibly seize farms, land, ranches, and coffee-producing farmlands, and impose tactics of terror on workers and farm owners (Business & Human Rights Resource

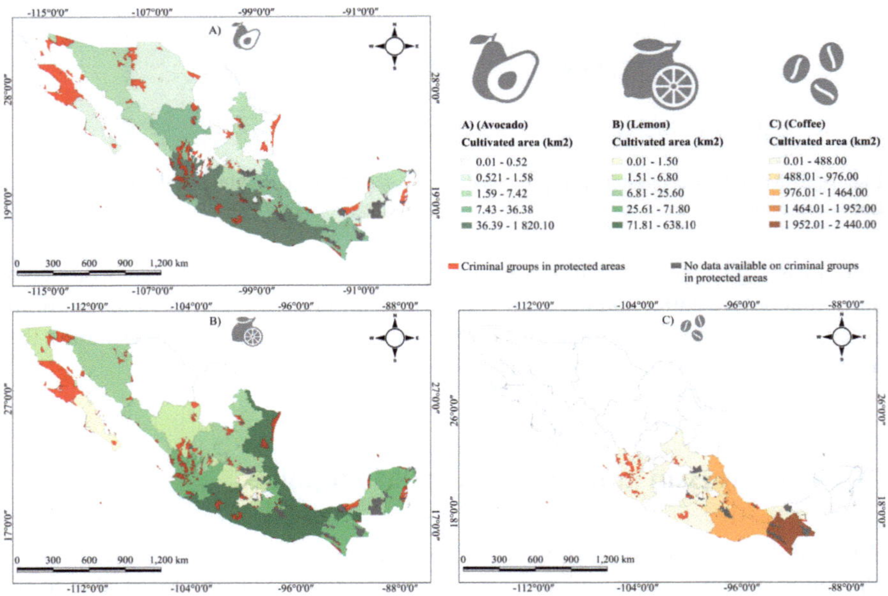

Fig. 10.3 Main regions of influence of organised crime in the agricultural sector in Mexico (avocado, lemon, and coffee) until 2025. Source: the authors with data from Secretaría de Agricultura y Desarrollo Rural (SADER) (2025) (Secretariat of Agriculture and Rural Development). Note: The geographical distribution is approximate due to limited access to official data

Centre, 2025). Forced labour, murders, and disappearances are criminal phenomena caused by drug trafficking and criminal groups in Mexico. Defending territories and human rights is punished with death, and the agricultural sector has been one of the most affected in the last decade.

Exploring the monopoly, social, economic and even environmental harms caused by criminal groups in Mexican agriculture would exceed the length of this chapter (and perhaps the entire book). However, it is necessary to highlight that environmental conservation activities in protected areas have been conducted in this same context for decades. The current conditions of public insecurity caused by criminal groups have impeded the defence and protection of the environmental, territorial, and human rights of rural and indigenous communities, but also the implementation and development of environmental conservation activities in protected areas.

In addition, attempts to defend the territory and environmental rights have led to the criminalisation, murder, and disappearance of people, mainly environmental defenders, academics, and public servants who have highlighted the negative impact of criminal groups on the environment, society, government, and public safety (see Carpio-Domínguez & Castro-Salazar, 2025; Centro Mexicano de Derecho Ambiental [CEMDA], 2023, 2024).

For green criminology, phenomena such as drug trafficking and organised crime cut across other forms of crime (Carpio-Domínguez et al., 2025; Anagnostou et al., 2025). In the current scenario in Mexico, environmental harms, human rights violations, and public safety intersect in the practices of drug trafficking and organised crime, but also in government neglect.

Carrabine et al. (2020) establish that one of the main lines of work for green criminology is to study the relationship between social inequalities and environmental harms, which are evident when analysing the impact of drug trafficking and criminal groups on indigenous and rural communities involved in drug cultivation and other illegal activities, largely due to the government's failure to meet their basic needs, leading them to find a livelihood in drug trafficking, but also in vulnerable populations such as public servants dedicated to environmental conservation in protected areas in contexts of violence.

Local Illegal Mining with International Implications

The convergence of organised crime and the illegal exploitation of natural resources in Mexico's protected areas has gone beyond illegal logging for sawmills, the diversion of streams and the creation of dams to store water for their drug crops, the use of natural spaces such as protected areas for drug production and as refuges and training camps. Recently, the involvement of criminal drug trafficking groups in illegal mining has also been documented.

Although the relationship between drug trafficking and the mining sector in Mexico is not new, given that in mining regions in northern Mexico there has been a tense but also collaborative relationship between large foreign mining companies and local criminal drug trafficking groups.

> When I joined the protected area team, I realised that the Canadian mining company was using sicarios to intimidate the indigenous communities and small settlements surrounding the mine […] when they refused to sell or rent their land, armed men would arrive and threaten them […] it was never reported in the news, especially because these communities were very remote and hidden in the mountains (04-Director of protected area, Sinaloa Cartel territory).

> Yes, the mine paid the cartel a protection fee to prevent attacks on trucks or shootings at planes and helicopters arriving with payroll money or transporting mine bosses. It was a Canadian mine (01-Environmental inspector, Sinaloa Cartel territory).

Among the policies implemented by Andrés Manuel López Obrador was a ban on granting further concessions to mining companies in the country, with the aim of protecting the national territory and guaranteeing environmental protection and investment in miners' wages (El Economista, 2019). However, 127 000 hectares were released for new mines in states such as Durango, Guanajuato, Guerrero, Jalisco, Michoacán, Nayarit, Nuevo León, Puebla, San Luis Potosí, Sinaloa, Sonora, and Zacatecas to foreign companies from Canada and the United States for the extraction of mainly gold, silver, copper, iron, uranium, lead, and zinc (see Lemus, 2019).

According to records, these policies of not approving concessions sought to reduce the environmental impact of the mining sector in protected areas. Even during López Obrador's six-year term, 1 609 mining concessions that had been granted in previous six-year terms were not authorised. These concessions were planned for 75% of Natural Resource Protection Areas (APRN), 63% of Biosphere Reserves (RB), 45% of Flora and Fauna Protection Areas (APFF), 22% of Sanctuaries (SANT), and 15% of National Parks (PN), prioritising the restitution of territories to indigenous communities (Angeles, 2021).

However, beyond political contradictions, organised crime began to see an unstable market with potential to exploit without intermediaries such as mining companies. During 2024 and 2025, violence intensified in indigenous and rural communities near mining deposits or areas with mining potential. Criminal groups such as the Sinaloa (Pacific) Cartel, through its factions such as Los Salazar, began to take over the gold in the Golden Triangle region, causing small mining companies to withdraw from the area and even launching attacks against communities using drones loaded with explosives (Mayorga, 2025).

The aim of these attacks has been to take control of mining deposits, causing the forced displacement of rural and indigenous communities and the murder of government representatives such as municipal presidents, as well as environmental and territory defenders (United Nations, 2023). Furthermore, this situation is exacerbated by corruption among government officials and members of criminal groups, which has left communities unprotected and without assessment of the potential environmental damage caused by the criminal groups' high-risk mining practices.

Meanwhile, criminal groups like the Jalisco Nueva Generacion Cartel have expanded their territory more than any other criminal group in the country and, in the process, have taken advantage of opportunities to generate economic income. Among the illegal activities recently documented in 2025 are incursions into illegal mining and mercury (Hg) trafficking, taking advantage of the lack of surveillance and the needs of rural and indigenous communities to obtain economic resources.

According to the report "Traffickers leave no stone unturned" by the Environmental Investigation Agency (EIA) (2025), with the increase in mercury prices since 2010, international demand for mercury, mainly in South American countries, has reactivated illegal mercury mining in Mexico.

From 2019 to 2025, the EIA (2025) estimates profits for the Jalisco Nueva Generacion Cartel of around $8 billion (US dollars) from mercury trafficking from Mexico, offering payments to rural and indigenous communities of $330 (US dollars) per kilogram of mercury. This practice violates the Minamata Convention on the regulation of mercury as a pollutant of wildlife, humans, and the environment.

With the entry into force of the Minamata Convention in Mexico in 2017, mercury production was reduced from 442 tonnes in 2017 to 0.5 tonnes in 2018, reaching zero for three consecutive years (2019–2021) and reporting the closure of mines that remained in operation in 2020. However, illegal and artisanal mercury mining has continued and even strengthened with the involvement of criminal groups, mainly in the Sierra Gorda Biosphere Reserve in Querétaro (see EIA, 2025).

Two main factors have been fundamental in organised crime's involvement in mercury mining and trafficking in the country. The first is that states such as Querétaro have been one of the main producers of mercury at the national level, with artisanal and illegal cinnabar (mercury sulphide, HgS) mines that have operated in the shadows despite government efforts to reduce and regulate mercury mining. The second is that criminal groups do not operate only at the national level. The Jalisco Nueva Generacion Cartel also operates in countries such as Bolivia, Peru, and Colombia, which are the main destination countries for mercury trafficked illegally by the Jalisco Nueva Generacion Cartel (Fig. 10.4).

The EIA (2025) reports that cinnabar (HgS) is mainly extracted in the Sierra Gorda Biosphere Reserve in the state of Querétaro, then processed in artisanal furnaces and stored. The mines and cinnabar processing furnaces are heavily guarded by members of the Jalisco New Generation Cartel. When a shipment of mercury is complete, it is transported to the ports of Manzanillo, Colima, and Veracruz, Veracruz, to be trafficked via two routes: the Pacific and the Caribbean.

Through bribes and document tampering, mercury is loaded and hidden in gravel shipments to be transported by sea to Arequipa in Peru via the Pacific Ocean. From there, it is trafficked by land to gold mines in Peru and crosses the border to gold mines in Bolivia. The other route is through the Caribbean. From the port of Veracruz, the mercury is also transported in ships loaded with gravel and unloaded on the coast of Colombia. It is then trafficked by land to Medellín and from there to the gold mines (EIA, 2025).

Sierra Gorda is a biosphere reserve, and various academic studies and media have reported the extraction of mercury using artisanal and improvised techniques for illegal sale (see El Cronista, 2025; Rodríguez, 2025; Hernández-Silva et al., 2012; Pérez-Campos & Herrera-Muñoz, 2013). This represents a problem of

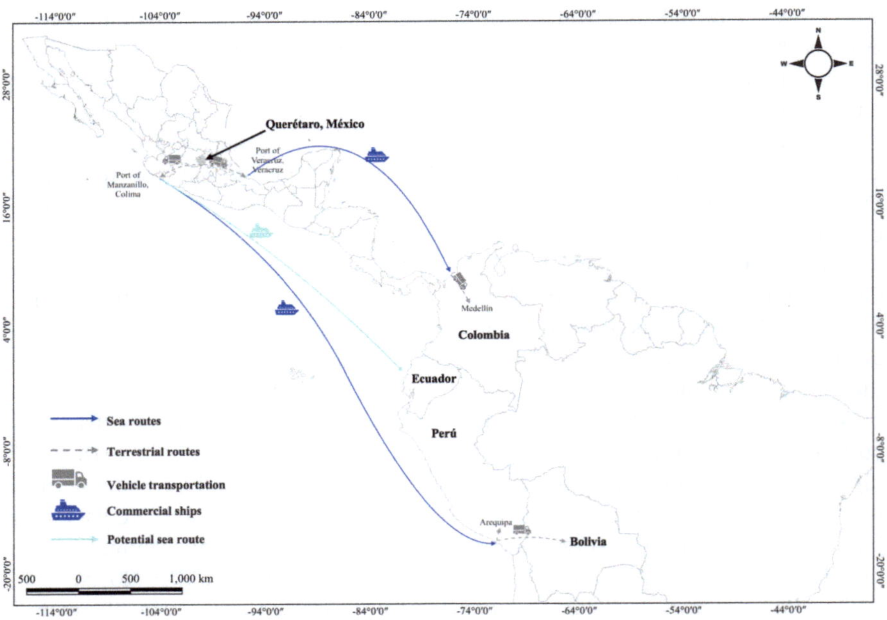

Fig. 10.4 Mercury (Hg) trafficking routes from Mexico to South America in 2025. Source: the authors with data from EIA (2025)

environmental, public health, administrative, and now also public safety concern (Jimémez, 2025; EIA, 2025).

This illegal activity has environmental impacts due to artisanal mercury extraction practices, which, according to studies conducted in cinnabar mines and mercury extraction sites, cause soil, water, and wildlife contamination, as well as serious impacts on the human health of mine workers who depend on illegal mining as their main source of livelihood. Furthermore, it causes environmental harm in destination locations in Peru, Colombia, and Bolivia; once the gold has been extracted, the remaining mercury is stored in facilities in rural areas without adequate waste management (EIA, 2025).

Human-to-Human Violence with Social and Environmental Harms in Protected Areas

Training Camps

The public security crisis in Mexico caused by organised crime has been evident not only in its ability to adapt and resist government efforts to regulate drug trafficking and dismantle criminal cells and cartels, but also by the government's inability not only to prevent and address these criminal phenomena, but also due to slowness, limited resources, bureaucratic processes, and political conflicts.

The use of natural spaces and their natural resources by organised crime is evident not only in the exploitation of these resources as a source of economic income, but also as strategic territories in which they exercise their dominance, and which are used for multiple criminal activities in addition to drug production and trafficking. Training camps are used for training, torture, and killing, and drug laboratories have been reported for decades in Mexico, but without official figures (e.g., BBC, 2025; Fernández, 2025; Ferri, 2025), and within protected areas, the records had gone unnoticed.

Mexico's protected areas, due to their natural characteristics and geographical location, are ideal territories for training camps. The remoteness of protected areas means that gunshots are not heard by other people, and together with drug laboratories, they represent one of the main criminal activities related to drug trafficking and criminal groups within and near protected areas.

According to the experiences of environmental inspectors and directors, organised crime training camps have been present in protected areas for more than 10 years and continue to be one of the main threats to environmental conservation activities within protected areas due to the risks involved in encountering and coming into contact with members of criminal groups in these spaces.

> When operatives enter the protected area to dismantle drug laboratories, they also find spent bullet casings and shooting targets, from which we can deduce that the drug laboratories also serve as training grounds for sicarios (hitmen) (08-Director of protected area, Sinaloa Cartel territory).

> About 12 years ago, we accidentally found an abandoned drug camp. There were tubs, buckets, machetes, cots, and blankets, but we also found a latrine and several shooting targets. I imagine that's where they practised (11-environmental inspector, territory of the Zetas Cartel).

> There are training camps everywhere. Years ago, there were also camps here in this protected area. I hadn't joined CONANP yet, but my colleagues told me all about what they had to go through in those days (16-Environmental inspector, CJNG Cartel territory).

Although organised crime and drug trafficking training camps have mainly been documented on ranches, farms, and other organised crime establishments (BBC, 2025; Fernández, 2025; Ferri, 2025), the existence of these training camps in protected areas endangers not only environmental conservation activities, but also the lives and physical integrity of CONANP environmental inspectors and directors.

With all the years of experience CONANP personnel have working in violent and unsafe environments within protected areas, the main strategy is non-confrontation due to the institutional policy of not carrying firearms, nor reporting them to the security forces due to the reprisals that such reports would entail, not only for CONANP personnel but also for the communities within and near the protected natural area. Therefore, they can only wait for drug destruction operations to be implemented within the protected area.

> We cannot report it. Recently, we saw smoke coming out of a ravine very close to the protected area. We know that it is a training camp for "la maña" [drug traffickers], but if we report it to public security, two things could happen: they could ignore us and/or they could tell them that we reported them and retaliate against us (14-Environmental inspector, territory of the Los Zetas Cartel).

It should be clarified that there are no forms of collaboration between CONANP personnel and criminal groups, nor have there ever been, nor is there any form of concealment of the illegal activities conducted by criminal groups. Rather, not reporting them is part of the adaptation and resilience strategies used to ensure environmental conservation and the development of communities within and near protected areas. Furthermore, CONANP is not a public security institution.

It is imperative not to confuse strategies for adapting to work in conditions of violence and insecurity with approval of illegal activities developed within protected areas. Rather, these strategies should be considered well-implemented responses to a context of violence that stems from structural failures and limitations (governmental, political, and even social) and that transcend the individual capacity of environmental inspectors and directors to report on illegal activities conducted by criminal groups and drug traffickers within protected areas.

Clandestine Graves

Among the criminal activities of organised crime and drug trafficking is the disappearance of persons, which implicitly involves homicide. During the Mexican War on Drugs, armed conflicts between criminal groups and state security forces led to homicides and disappearances, as did subsequent conflicts between groups vying for control of larger territories and divisions within each cartel.

However, as an alternative to narco-terrorism through the display of dismembered or decapitated bodies of rival members of criminal groups in public squares, roads, or avenues in cities (Arteaga-Botello, 2009; Chávez-Jiménez & Ramírez-Sánchez, 2024; Reig, 2011), there was another practice that has continued to this day: the disappearance of persons and clandestine graves.

The number of missing persons in Mexico stands at 133,602 people missing and not located until November 2025, with the states of Jalisco, Tamaulipas, and the Estado de México having the highest numbers (Comisión Nacional de Búsqueda, 2025) (National Commission for the Search for Missing Persons). This criminal phenomenon has serious social, legal, victimological, and public safety implications. The disappearance of persons includes not only homicide, but also burying, burning, and/or destroying the victim's body so that it cannot be found, causing psychological, economic, and social harms to families (see Bernal-Ramírez, 2025; Lorusso, 2021; Vicente-Ovalle, 2021; Villarreal-Martínez, 2016, 2022).

The disappearance of people has resulted in the existence of more than 5696 clandestine graves in more than 570 municipalities in Mexico as of 2023 (Tzuc, 2023), although it is highly plausible that this figure is higher. Furthermore, transcending the image of clandestine graves as improvised tombs in which people are buried, the concept of extermination camps, also known as extermination kitchens (see Herrera-Buenrostro & Torres-Alfaro, 2024; Márquez-Verduzco, 2022), confer a dimension beyond "removing people from the path" through homicide.

The extermination camps and kitchens focus on ensuring that missing persons are never found or identified, by destroying the bodies through burning and grinding the bones into small, unrecognisable fragments, which are then buried or scattered over a large area (Herrera-Buenrostro & Torres-Alfaro, 2024).

This phenomenon is also related to protected areas in Mexico because these strategic territories over which they exercise control are also used to disappear people, and where there are clandestine graves in which the bodies of innocent citizens (children, young people, women, and men), businesspeople, members of rival criminal groups, and public servants lie buried or destroyed, who have not been found by their families and without access to justice (see Osorio-González, 2023) (Fig. 10.5).

An analysis of data from Quinto Elemento Lab (Tzuc, 2023) on clandestine graves in Mexico identified 106 clandestine graves within Mexico's protected areas as of 2023. The states of Veracruz ($n = 58$) and San Luis Potosí ($n = 14$) have the highest number of clandestine graves within Mexico's protected areas, and both states have a presence of the Jalisco Nueva Generacion Cartel.

This coincides with the experiences of environmental inspectors and directors of protected areas, who highlight the presence of organised crime vehicles within protected areas, as well as the processing and investigation of clandestine graves by public security institutions and the presence of training camps within protected areas.

> So far this year [2024], missing persons search groups have found two clandestine graves in the protected area, on the periphery, near the town (12-Environmental inspector, territory of the La Familia Cartel).

> We have seen trucks with armed individuals [sicarios] entering some zones of the protected area, and they have seen us, but they do not care. They act as if they do not see us but knowing about all the graves that have been found nearby, we have no doubt that they are using the protected area to bury bodies (15-Environmental inspector, CJNG Cartel territory).

> I don't want to get involved in these things, but I would be lying if I said I didn't think there were clandestine graves in the protected area (16-Environmental inspector, CJNG Cartel territory).

> Those are serious words [dangerous topics]. I don't want to say anything that will get me into trouble, but what I can tell you is that planting drugs and things like that in the protected area is one thing, but using the protected area to kill and bury people is something completely different (14-Environmental inspector, Los Zetas Cartel territory).

The clandestine graves found in Mexico and in protected areas date back to different periods of time, so this is not a recent criminal practice. However, it is a criminal phenomenon present in protected areas and adds a new dimension to the complexity and risk of environmental conservation activities in protected areas in contexts of violence.

It is evident that environmental inspectors and directors of protected areas are cautious when discussing the disappearance of people and clandestine graves within

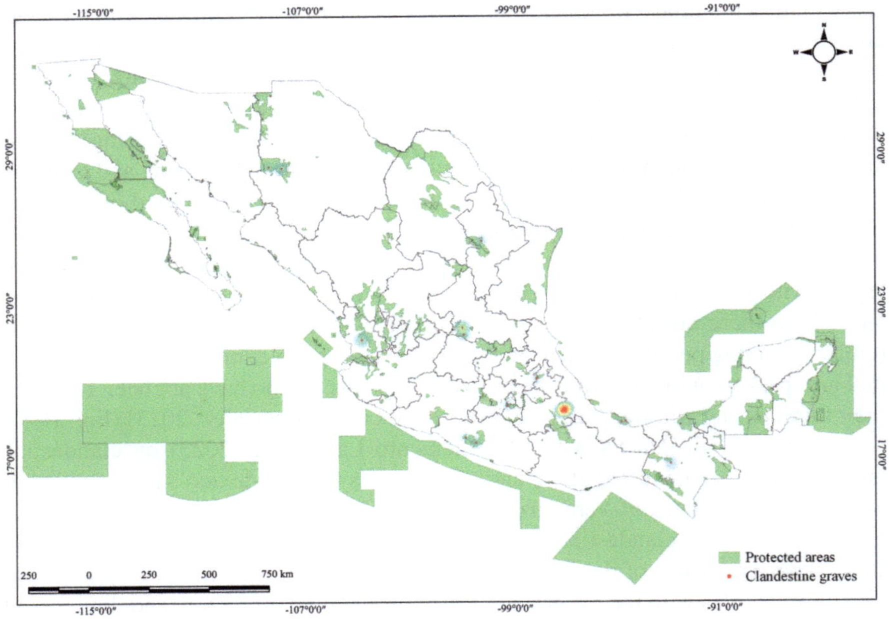

Fig. 10.5 Clandestine graves reported in (federal) protected areas in Mexico until 2025. Source: the authors with data from Quinto Elemento Lab (Tzuc, 2023) and CONANP (2025). Note: It is highly possible that the map does not show the total of clandestine graves in all protected areas, as it only shows federally protected areas

protected areas, attributing (albeit not explicitly) a higher level of danger than the mere presence of these groups within and near Mexico's protected areas.

Further studies are needed on clandestine graves and training and extermination camps in protected areas. However, this section helps to highlight and document the intersection of these criminal public security phenomena with environmental issues in Mexico (Carpio-Domínguez, 2021). Furthermore, this study only considered the intersection of clandestine graves in protected areas under federal jurisdiction, so future studies are needed that also consider protected areas under each state jurisdiction.

Drug Traffickers as Environmental Conservationists?

Unapologetically, the presence of criminal groups in Mexico's protected areas has also changed other criminal practices that posed serious problems for environmental inspectors and directors within protected areas. In 2019, the Facebook page HERP.mx (2019) referred to drug traffickers in protected areas and other wilderness areas as "intermittent conservationists" because the presence of these criminal groups limits poaching, the extraction of flora from the wild, and uncontrolled livestock grazing in natural areas.

The study that preceded this book (Carpio-Domínguez, 2021) documented the positive impact on environmental conservation of the presence of criminal groups within protected areas. Expanding on that study, it was identified that incursions by individuals into protected areas pose a threat to the illegal activities conducted by criminal groups, such as drug cultivation, trafficking routes, and training camps. Therefore, criminal group members react with mistrust and aggression towards external people within the area.

The absence of external people within protected areas has had positive impacts in certain zones within the areas, allowing for the recovery of vegetation cover, a reduction in the number of small dams built as watering holes for livestock, and a decrease in practices such as illegal hunting, illegal fishing, and the extraction of wildlife (flora and fauna) due to the fear of encountering drug traffickers within the protected area.

> Illegal hunting has decreased due to the presence of these groups (drug traffickers) in the area (02-Environmental inspector, territory of the Sinaloa Cartel).

> People no longer go into the forest to hunt animals. They used to hunt deer, wild boar and turkeys, but now they don't go in anymore. What we have noticed is that the forest remains more or less intact. No one cuts down trees, they don't even go near the forest… mainly because they are afraid (09-Director of protected area, Gulf Cartel territory).

> Illegal fishing has decreased. People who used to come at night or in the early morning to fish illegally no longer come, because that is when the drug traffickers pass through with their drugs, and if they see them fishing, things will go badly for them (10-environmental inspector, Tijuana Cartel territory).

> One thing we have noticed is that we no longer have illegal hunters, not even for personal consumption. When these people [drug traffickers] entered the protected area, the local people gradually stopped going into the forest […]. Yes, they used to go in and hunt deer and turkeys, they even brought back parrots as pets, but they stopped doing so for fear of encountering the mañosos [drug traffickers] (16-Environmental inspector, CJNG Cartel territory).

This has also been reported in countries such as Colombia, where the presence of armed groups in wilderness areas has led to the conservation of large extensions of natural territory, which have remained untouched by large development projects (Morales, 2017). In the case of Mexico, it is evident that the presence of criminal groups in protected areas has led to environmental conservation due to criminal environmental practices by private individuals. However, these unintended results of the presence of these criminal groups in protected areas do not compensate for the social, economic, and environmental harms they cause, as well as the institutional limitations on CONANP personnel to conduct environmental conservation activities in a professional manner and to promote the social development of rural and indigenous communities within and near protected areas.

This ambivalence of force in Mexico's protected areas not only highlights the need to strengthen the capacities of public security and environmental institutions but also reveals that criminal groups exercise control over territories, government institutions (public security and environmental), rural, indigenous, and urban communities, as well as over nature. This control exercised by criminal groups in almost all spheres of human life in Mexico can also be seen and made visible in protected areas.

Conclusions

Protected areas are territories where public safety issues intersect with environmental conservation. Protected areas should be the bastion of environmental conservation in Mexico. However, as this study shows, these are territories where public safety issues, institutional limitations, physical vulnerability, disappearances, threats to biodiversity and ecosystems, foreign and local extractivism, organised crime, and government neglect all coincide.

However, CONANP personnel also demonstrate a dedication to service that transcends all the issues highlighted in this book. Mexico's protected areas are safeguarded and managed by people (civil servants) who see protected areas as a means of guaranteeing quality of life for future generations and protecting human and non-human life.

Even with limited access to official data and phenomena occurring in real time, this chapter demonstrates that criminal activities and new illicit markets in Mexico's protected areas not only jeopardise environmental conservation but also undermine homeland security (not from the outside, but from within). Public safety issues continue to threaten the environment in Mexico, and environmental conservation has relied on public servants who demonstrate an unwavering commitment to environmental conservation above personal or economic interests.

Mexico's protected areas are valuable not only for their natural resources. These spaces are also home to the cultural and historical roots of modern Mexican society. Their conservation helps to protect the cultural legacy and heritage that is evident in wild animals, plants, ecosystems, and habitats, as well as landscapes, rainfall, trees, deserts, and mangroves, coastlines and the depths of the sea, which form part of the natural diversity that makes Mexico one of the most megadiverse countries in the world, but also unique in the ways in which humanity relates to nature.

Finally, the protection of Mexico's protected areas should not only be the responsibility of CONANP personnel, but also of each person individually and collectively. Environmental governance activities involving all actors in social life—government, citizens, academia, and the private sector—should continue in order to promote shared responsibility for the protection of Mexico's natural heritage.

References

Anagnostou, M., Doberstein, B., Armitage, D., Stoett, P., & Glasson, A. (2025). Disentangling and demystifying converging crimes and illegal wildlife trade in South Africa, Hong Kong, and Canada. *Journal of Economic Criminology, 10*, 100196. https://doi.org/10.1016/j.jeconc.2025.100196

Angeles, J. (2021). *Hemos recuperado millones de hectáreas cancelando concesiones mineras: AMLO*. RegeneracionMX (May 20, 2021). Available on https://regeneracion.mx/hemos-recuperado-millones-de-hectareas-cancelando-concesiones-mineras-amlo/?amp

Arteaga Botello, N. (2009). Decapitaciones y mutilaciones en el México contemporáneo. *Espacio Abierto, 18*(3), 463–486. https://www.redalyc.org/pdf/122/12211825004.pdf

Balboni, C., Berman, A., Burgess, R., & Olken, B. (2023). The economics of tropical deforestation. *Annual Review of Economics, 15*, 723–754. https://doi.org/10.1146/annurev-economics-090622-024705

BBC. (2025). *Qué se sabe del Rancho Izaguirre, el lugar de "reclutamiento y exterminio" del crimen organizado en México donde se hallaron fosas clandestinas y restos óseos*. BBC News Mundo (March 13, 2025). Availabe on https://www.bbc.com/mundo/articles/c39v44pp1x9o

Bernal-Ramírez, M. (2025). Subjetividad del muerto viviente. El fenómeno de desaparición de personas en México y la crisis de fentanilo en los E.U.A. *Norteamérica, 20*(1), 1–24. https://doi.org/10.22201/cisan.24487228e.2025.1.720

Buscaglia, E. (2015). *Lavado de dinero y corrupción política: el arte de la delincuencia organizada internacional*. Penguin Random House Grupo Editorial.

Business & Human Rights Resource Centre. (2025). *México: Crimen organizado esclaviza a personas trabajadoras en fincas cafetaleras de Chiapas y vende a grandes comercializadoras*. Business & Human Rights Resource Centre (February 9, 2025). Available on https://www.business-humanrights.org/en/latest-news/méxico-fincas-cafetaleras-de-chiapas-tomadas-por-el-crimen-organizado-para-explotar-y-amenazar-a-personas-trabajadoras/

Cameron, O. (2019). *"Narco-Deforestation" in Central American Protected Areas: Developing A Spatiotemporal Proxy Database for Drug-Trafficking Activities from Media Reporting*. Thesis: Oregon State University. Available on https://ir.library.oregonstate.edu/concern/honors_college_theses/fn107447s

Cano, J. (2024). *Narcomapa 2024: qué cárteles tienen mayor presencia en México*. Infobae (23 de abril de 2024). México. https://www.infobae.com/mexico/2024/04/23/narcomapa-2024-que-carteles-tienen-mayor-presencia-en-mexico/

Carbajal, B. (2023). *Corrupción y 'narco' asedian el millonario negocio del aguacate*. La Jornada (February 11, 2025). Available on https://www.jornada.com.mx/noticia/2023/02/11/reportaje/corrupcion-y-narco-asedian-el-millonario-negocio-del-aguacate-8989

Carpio-Domínguez, J. (2021). Crimen Organizado (narcotráfico) y conservación ambiental: el tema pendiente de la seguridad pública en México. *Revista CS, 33*, 237–274. https://doi.org/10.18046/recs.i33.4076

Carpio-Domínguez, J. (2024). The harms and crimes of logging and deforestation. *Oxford Research Encyclopedia of Criminology*. https://doi.org/10.1093/acrefore/9780190264079.013.769

Carpio-Domínguez, J., & Castro-Salazar, J. (2025). Researching environmental harms in violent environments of the global south. In R. P. Cavalcanti, D. S. Fonseca, V. Vegh Weis, K. Carrington, R. Hogg, & J. Scott (Eds.), *The Palgrave handbook of criminology and the global south*. Palgrave Macmillan. https://doi.org/10.1007/978-3-031-74932-2_115-1

Carpio-Domínguez, J., Cervantes-Niño, J., Castro-Salazar, J., & Mendezcarlo-Silva, V. (2025). Policing wildlife trafficking in northeastern Mexico: The case of Tamaulipas in 2023-2024. *Frontiers in Conservation Sciences, 6*, 1488500. https://doi.org/10.3389/fcosc.2025.1488500

Carrabine, E., Cox, A., Cox, P., Crowhurst, I., de Ronco, A., Fussey, P., Carrabine, E., Cox, A., Cox, P., Crowhurst, I., di Ronco, A., Fussey, P., Sergi, A., South, N., Thiel, D., & y Turton, J. (2020). *Criminology: A sociological introduction*. Routledge. https://doi.org/10.4324/9781315123509

Centro Mexicano de Derecho Ambiental (CEMDA). (2023). *Informe sobre la situación de las personas y comunidades defensoras de los derechos humanos ambientales en México 2023*. CEMDA. https://cemda.org.mx/informe-sobre-la-situacion-de-las-personas-y-comunidades-defensoras-de-los-derechos-humanos-ambientales-en-mexico-2023/

Centro Mexicano de Derecho Ambiental (CEMDA). (2024). *Informe sobre la situación de las personas y comunidades defensoras de los derechos humanos ambientales en México 2024*. CEMDA. https://cemda.org.mx/wp-content/uploads/2025/04/INFORMEDEFENSORES_2024_WEB.pdf

Chávez-Jiménez, D., & Ramírez-Sánchez, R. (2024). Las violencias en México, fenómeno en expansión (2006-2022). *Inventio, 19*(48), 1–9. https://doi.org/10.30973/2023.19.48/2

Cheeseman, K. (2024). The environmental impacts of illicit drug production. *The Institute of Development Studies and Partner Organisations*. Available on: https://hdl.handle.net/20.500.12413/18339

Comisión Nacional de Áreas Naturales Protegidas (CONANP). (2025). *Sistema de Información, Monitoreo y Evaluación para la Conservación (SIMEC)*. Gobierno de México. Available on https://simec.conanp.gob.mx/

Comisión Nacional de Búsqueda. (2025). *Versión estadística del Registro Nacional de Personas Desaparecidas y No Localizadas* (17 de noviembre de 2025). Gobierno de México. Available on https://versionpublicarnpdno.segob.gob.mx/Dashboard/ContextoGeneral

Comisión Nacional Forestal (CONAFOR). (2024). *Sistema Nacional de Información Forestal*. Gobierno de México. Available on https://snif.cnf.gob.mx

Cotrina-Sánchez, A., Bandopadhyay, S., Rojas Briceño, N., Banerjee, P., & Torres Guzmán, M. (2021). Peruvian Amazon disappearing: Transformation of protected areas during the last two decades (2001–2019) and potential future deforestation modelling using cloud computing and MaxEnt approach. *Journal for Nature Conservation, 64*, 126081. https://doi.org/10.1016/j.jnc.2021.126081

Cumplido-Tercero, M. (2015). *Evolución del crimen organizado en México. Más allá del narcotráfico*. Instituto Español de Estudios Estratégicos. Documento Marco. Available on https://dialnet.unirioja.es/servlet/articulo?codigo=7687260

El Cronista. (2025). *La droga secreta que tienen los carteles y es peor que el fentanilo y la cocaína: está prohibida en más de 100 países*. El Cronista (July 28, 2025). Available on https://www.cronista.com/mexico/actualidad-mx/la-droga-secreta-que-tienen-los-carteles-y-es-peor-que-el-fentanilo-y-la-cocaina-esta-prohibida-en-mas-de-100-paises/

El Economista. (2019). *AMLO advierte que no entregará más concesiones a empresas mineras*. Notimex (August 11, 2019). Available on https://www.eleconomista.com.mx/politica/AMLO-advierte-que-no-entregara-mas-concesiones-a-empresas-mineras-20190811-0012.html

Environmental Investigation Agency (EIA). (2025). *Traffickers leave no stone unturned*. United States: EIA US Government. Available on https://eia.org/wp-content/uploads/2025/07/EIA_US_Mercury_Smuggling_report_0725_FINAL.pdf

España-Boquera, M., & Champo-Jiménez, O. (2016). Proceso de deforestación en el municipio de Cherán, Michoacán, México (2006-2012). *Madera y bosques, 22*(1), 141–153. http://www.scielo.org.mx/scielo.php?script=sci_arttext&pid=S1405-04712016000100141&lng=es&tlng=es

Fernández, M. (2025). *Campo de entrenamiento en México expone los horrores del reclutamiento forzado del CJNG*. Insight Crime (March 25, 2025). Availabe on https://insightcrime.org/es/noticias/campo-exterminio-mexico-expone-horrores-reclutamiento-forzado-cjng/

Ferri, P. (2025). *Reclutados por el cartel en el rancho La Vega: "Me ofrecieron trabajo por Facebook, pero luego me dijeron que sería sicario o me descuartizaban"*. El País (September 21, 2025). Available on https://elpais.com/mexico/2025-09-21/reclutados-por-el-cartel-en-el-rancho-la-vega-me-ofrecieron-trabajo-por-facebook-pero-luego-me-dijeron-que-seria-sicario-o-me-descuartizaban.html

Food and Agriculture Organization of the United Nations. (2010). *Terms and definitions: Global forest resource assessment*. FAO. Available on http://www.fao.org/docrep/014/am665e/am665e00.pdf

Fuentes-Díaz, A. (2015). Narcotráfico y autodefensa comunitaria en "Tierra Caliente", Michoacán. *México. CienciaUAT, 10*(1), 68–82. http://www.scielo.org.mx/scielo.php?script=sci_arttext&pid=S2007-78582015000200068&lng=es&tlng=es

Hernández-Silva, G., García-Martínez, R., Solís-Valdez, S., Martínez-Trinidad, S., Mercado-Sotelo, I., Ramírez-Islas, M., Scharek, P., & Solorio-Munguía, G. (2012). Presencia del Hg total en una relación suelo-planta-atmósfera al sur de la Sierra Gorda de Querétaro, México. *TIP, 15*(1), 05–15. http://www.scielo.org.mx/scielo.php?script=sci_arttext&pid=S1405-888X2012000100001&lng=es

Herp.mx. (2019). Narcos, los conservacionistas intermitentes. HERP.mx (May 20, 2019). Available on https://es-la.facebook.com/herpmx/posts/narcos-the-fickle-conservationistsdrug-cultivation-isa-double-edged-sword-for-m/2305989742794969/

Herrera-Buenrostro, K., & Torres-Alfaro, D. (2024). Desafíos actuales ante la crisis forense y la presencia del crimen organizado en México. La intervención del arqueólogo forense en fosas clandestinas y campos de exterminio. *Gaceta internacional de ciencias forenses, 53*, 13–26. https://www.uv.es/gicf/4A1_Torres_GICF_53.pdf

Jimémez, E. (2025). *El CJNG expande su dominio en Michoacán y comete delitos ambientales en zonas protegidas*. Infobae (October 27, 2025). Available on https://www.infobae.com/mexico/2023/10/28/el-cjng-expande-su-dominio-en-michoacan-y-comete-delitos-ambientales-en-zonas-protegidas/

Lemus, J. (2019). *AMLO y sus contradicciones: anuncia que no habrá más concesiones mineras, pero libera 127 mil hectáreas para nuevas minas*. REMA (March 24, 2019). Available on https://www.remamx.org/2019/03/amlo-y-sus-contradicciones-anuncia-que-no-habra-mas-concesiones-mineras-pero-libera-127-mil-hectareas-para-nuevas-minas/

Lorusso, F. (2021). Una discusión sobre el concepto de fosa clandestina y el contexto mexicano. El caso de Guanajuato. *Historia y grafía, 56*, 129–170. https://doi.org/10.48102/hyg.vi56.354

Maguey, H., & Francisco, D. (2022). *La mayoría de los incendios, provocados por actividades ilícitas*. Gaceta UNAM (April 4, 2022). Mexico. Available on https://www.gaceta.unam.mx/la-mayoria-de-los-incendios-provocados-por-actividades-ilicitas/

Márquez-Verduzco, D. (2022). Necropoder y subjetividad: la desaparición de personas en el norte de Veracruz, México. *Andamios, 19*(50), 135–165. https://doi.org/10.29092/uacm.v19i50.947

Mayorga, P. (2025). *En busca de oro, el narco bombardea con drones para desplazar a indígenas tarahumaras*. Proceso (July 3, 2025). Available on https://www.proceso.com.mx/nacional/2025/7/3/en-busca-de-oro-el-narco-bombardea-con-drones-para-desplazar-indigenas-tarahumaras-354144.html

McSweeney, K., Nielsen, E., Taylor, M., Wrathall, D., Pearson, Z., Wang, O., & Plumb, S. (2014). Drug policy as conservation policy: Narco-deforestation. *Science, 343*(6170), 489–490. https://doi.org/10.1126/science.1244082

Morales, L. (2017). *La paz y la protección ambiental en Colombia: propuestas para un desarrollo rural sostenible*. El Diálogo. Liderazgo par las Américas. Available on https://www.thedialogue.org/wp-content/uploads/2017/01/Envt-Colombia-Esp_Web-Res_Final-for-web.pdf

Morales-Gámez, L., & Ortega-Ramírez, A. (2024). Security and interdependence between Mexico and the United States (2019-2023): How do they avoid declaring drug traffickers as terrorists? *Norteamerica, 19*(2), 219–240. https://doi.org/10.22201/cisan.24487228e.2024.2.656

Moreno, A. (2024). Facetas de la polarización política en México. *Foro internacional, 64*(2), 189–238. https://doi.org/10.24201/fi.v64i2.3064

Mujica, J., Zevallos-Trigoso, N., & Campos-Vásquez, C. (2024). Vigilar y performar. La cadena de suministro de la madera ilegal en Perú. *URVIO Revista Latinoamericana De Estudios De Seguridad, 38*, 49–68. https://doi.org/10.17141/urvio.38.2024.6093

Osorio-González, A. (2023). *Un paraiso protegido convertido en fosa clandestina*. A dónde van los desparecidos (December 7, 2023). Available on https://adondevanlosdesaparecidos.org/2023/12/07/un-paraiso-protegido-transformado-en-fosa-clandestina/

Pérez-Campos, E., & Herrera-Muñoz, A. (2013). Minas y mineros: presencia de metales en sedimentos y restos humanos al sur de la Sierra Gorda de Querétaro en México. *Chungará (Arica), 45*(1), 161–176. https://doi.org/10.4067/S0717-73562013000100008

Reig, R. (2011). Periodismo y muerte: bases teóricas y psicosociales, y el caso de México. *Razón y Palabra. Primera Revista Electrónica en América Latina Especializada en Comunicación, 74*, 1–30. Available on https://web.archive.org/web/20180428121136id_/http://www.razonypalabra.org.mx/N/N74/VARIA74/32ReigV74.pdf

Rivera Magos, S., & González Pureco, G. (2024). Populismo, desinformación y polarización política en la comunicación en redes sociales de los presidentes populistas latinoamericanos. *Revista Mexicana De Opinión Pública, 36*, 79–107. https://doi.org/10.22201/fcpys.24484911e.2024.36.86828

Roa, J. (2020). Presencia de grupos criminales en México (2018-2019). Programa de Política de Drogas. https://www.politicadedrogas.org/PPD/index.php/observatorio/metaDescargables/id/137.html

Rodríguez, A. (2025). *Tráfico de mercurio para las minas de oro: el nuevo y millonario 'negocio' del Cartel Jalisco Nueva Generación.* El País (August 3, 2025). Available on https://elpais.com/mexico/2025-08-03/trafico-de-mercurio-para-las-minas-de-oro-el-nuevo-y-millonario-negocio-del-cartel-jalisco-nueva-generacion.html

Santos-Duarte, H., Schmidt, A., & Wahl, S. (2021). Addressing coca-related deforestation in Colombia: A call for aligning drug and environmental policies for sustainable development. *Journal of Illicit Economies and Development, 3*(1), 10–21. https://doi.org/10.31389/jied.79

Sarsfield, R. (2023). Entre el pueblo bueno y la élite corrupta. Narrativa populista y polarización afectiva en las redes sociales en México. *Revista mexicana de opinión pública, 35*, 13–34. https://doi.org/10.22201/fcpys.24484911e.2023.35.85518

Secretaría de Agricultura y Desarrollo Rural (SADER). (2025). *Anuario Estadístico de la Producción Agrícola.* Gobierno de México. Availabe on https://nube.agricultura.gob.mx/cierre_agricola/

Tzuc, E. (2023). *México rebasa las 5 mil 600 fosas clandestinas.* Quinto Elemento Lab (October 9, 2025). Available on https://quintoelab.org/project/mexico-rebasa-cinco-mil-fosas-clandestinas

United Nations. (2023). *Protegiendo a las comunidades tarahumaras en México.* United Nation News (Novembrer 6, 2023). https://news.un.org/es/story/2023/11/1525422

Valdés, G. (2013). *Historia del narcotráfico en México.* Editorial Aguilar.

Valdéz, A., & Sánchez, M. (2020). "Oro verde", educación e inseguridad en ciudades medias: Ciudad Guzmán y Uruapan, México. *Revista de Ciencias Sociales (Ve), 26*(2), 414–425. https://doi.org/10.31876/rcs.v26i0.34136

Van Solinge, T. (2010a). Deforestation crimes and conflicts in the Amazon. *Critical Criminology, 18*(4), 263–277. https://doi.org/10.1007/s10612-010-9120-x

Van Solinge, T. (2010b). Equatorial deforestation as a harmful practice and criminological issue. In R. White (Ed.), *Global environmental harm: Criminological perspectives* (pp. 20–36). Willan.

Van Solinge, T. (2014). Researching illegal logging and deforestation. *International Journal for Crime, Justice and Social Democracy, 3*(2), 35–48. https://doi.org/10.5204/ijcjsd.v3i2.179

Vázquez-Valdéz, J. (2018). Economía del narco: prohibicionismo, violencias sistémicas y capital criminal. *Caleidoscopio-Revista Semestral De Ciencias Sociales Y Humanidades, 22*(38), 105–130. https://doi.org/10.33064/38crscsh917

Vicente-Ovalle, C. (2021). Desapariciones en México: la emergencia de un campo. *Historia y grafía, 56*, 53–87. https://doi.org/10.48102/hyg.vi56.353

Villarreal-Martínez, M. (2016). Los colectivos de familiares de personas desaparecidas y la procuración de justicia. *Intersticios sociales, 11*, 1–28. http://www.scielo.org.mx/scielo.php?script=sci_arttext&pid=S2007-49642016000100007&lng=es&tlng=es

Villarreal-Martínez, M. (2022). Respuestas ciudadanas ante la desaparición de personas en México (2000-2013). *Espacios Públicos, 17*(39), 105–135. https://espaciospublicos.uaemex.mx/article/view/19444

Index

J. L. Carpio-Domínguez et al., *Drug Trafficking and Protected Areas in Mexico*, https://doi.org/10.1007/978-3-032-18218-0

MIX
Papier aus verantwortungsvollen Quellen
Paper from responsible sources
FSC® C105338

If you have any concerns about our products,
you can contact us on
ProductSafety@springernature.com

In case Publisher is established outside the EU,
the EU authorized representative is:
Springer Nature Customer Service Center GmbH
Europaplatz 3, 69115 Heidelberg, Germany

Printed by Libri Plureos GmbH
in Hamburg, Germany